DECODING ELON MUSK'S SECRET MASTER PLANS

Why Electric Vehicles and Solar Are A Winning Financial Strategy

NEO TRINITY

Author and Executive Producer: Neo Trinity

Book cover art design: Neo Trinity

Book cover image: © 2022 Immortal Magnum Opus – "Elon Musk Multidimensional Dynamic Facets Transcending Space and Time" – Copyright owned by Solar Multiplier, Inc.

All trademarks herein are respectively owned by Solar Multiplier, Inc. or Energy Trinity, Inc.

First Edition – September 2022

www.energytrinity.com

ISBN: 979-8-9870283-0-8

Dedication

To my wife Debbie and children Garrett and Kristen. I am so truly blessed and fortunate to have such a loving and wonderful family. All the many adventures and journeys we have experienced together as a family traveling around the world have been totally epic! I am deeply humbled and grateful that our family always brings out the very best in each other. Thank you for always bringing out the best in me as a mentor, a leader, a husband, and as a father. Each one of you, along with the collective experiences we have shared together, have been love and joy multipliers in my life. My life has only gotten better and brighter the very moment each one of you joyfully entered into it. Thank you for giving, showing, and sharing with me a life of love without limits and a life lived without regrets.

NEO

Disclaimer and Nonliability

The author used best efforts in preparing this publication. No representations or warranties with respect to the accuracy or completeness of the contents are made. Any implied savings or warranties are disclaimed. No warranty may be created or extended by sales representatives or written sales materials from this book.

The strategies and advice contained herein may not be suitable for your situation. Actual results may materially occur and be different from forecasted projections. You should consult with a professional advisor where appropriate.

The information in this publication is not intended or implied to be a substitute for professional investing advice inclusive of electric vehicles or professional solar energy advice. This publication makes no representations and assumes no responsibility for the accuracy of information contained or available through this publication. Additionally, such information is subject to change without notice.

This publication does not constitute a recommendation, endorsement, or representation about the accuracy of products, services, opinions, professionals, or other information that may be contained in this book. The author and information contained herein are not liable for any damages arising herefrom. Additionally, the author specifically disclaims any liability that might be incurred from the use or application of the contents in this book.

TABLE OF CONTENTS

1

GETTING THREE TESLA MODEL 3s AND SOLAR COSTING $190,000 – FOR FREE!

"That free fusion reactor in the sky conveniently converts ~4 million tons of mass into energy every second. We just need to catch an extremely tiny amount of it to power all of civilization." – Elon Musk

People are baffled when I tell them that my family's three Tesla Model 3 Electric Vehicles (EV) are free. They laugh and dismiss me in total disbelief. How can $150,000+ USD in combined Tesla purchases be free? But wait a second, not only are our three Tesla Model 3s free, we installed a Solar Energy System at a cost of $38,000, and that too, is also free. Now we are up to almost $190,000 USD in combined Tesla and Solar Energy cost outlays. How can this be? The answer is Solar. Yes, SOLAR!

Don't believe it? Consumers who already own solar most likely understand the concept of cost recovery or payback period on their solar investment. These consumers most likely also understand the concept of cost savings. Payback represents the period of time it takes to recover the cost of an investment. Cost savings represents the reduction or elimination of expenses. It's been financially proven (when done correctly) that by strategically investing in a Solar Energy System, the financial cost outlay will be recovered over time, typically 7 – 15 years, due to the elimination of electricity costs. Once a Solar Energy System investment has been fully cost recovered, it then becomes a cost savings by eliminating future electricity expense.

What no one realizes is that by pairing an Electric Vehicle with a Solar Energy System, it becomes a financial cost savings multiplier, allowing one to recoup not only their solar investment costs but also their Electric Vehicle costs. In most cases, a majority of it, and in some cases, when your Solar Energy System is spec'd out, such as our family's, all of it is recoverable. By recovering our $190,000 USD initial cost outlays, our three Tesla's and Solar Energy System becomes free.

Still don't believe it? I will demonstrate how we spent $190,000 on three Tesla Model 3s and a Solar Energy System and how we are recovering 100% of our investment, thereby making them ALL FREE! Read on as each element in this book reveals step by step why Electric Vehicles and Solar are a winning financial strategy.

I financially decoded two key aspects within Elon Musk's Secret Master Plans. These two key aspects are:

1) Being energy positive.
2) Empowering the individual as their own utility.

Being "Energy Positive" was revealed in 2006 – **The Secret**

2

Tesla Motors Master Plan. Musk stated, *"I should mention that Tesla Motors will be co-marketing sustainable energy products from other companies along with the car. For example, among other choices, we will be offering a modestly sized and priced solar panel from SolarCity, a photovoltaics company (where I am also the principal financier). This system can be installed on your roof in an out of the way location, because of its small size, or set up as a carport and will generate about 50 miles per day of electricity.*

If you travel less than 350 miles per week, you will therefore be "energy positive" with respect to your personal transportation. This is a step beyond conserving or even nullifying your use of energy for transport – you will actually be putting more energy back into the system than you consume in transportation!"

"Empowering the individual as their own utility" was revealed in 2016 – **Master Plan, Part Deux** under – Integrate Energy Generation and Storage. Musk stated, *"Create a smoothly integrated and beautiful solar-roof-with-battery product that just works, empowering the individual as their own utility, and then scale that throughout the world."*

Both of these two components stood out to me. What happens to an individual who becomes energy positive by being empowered as their own utility?

I approached these ideas from a purely financial perspective. My family and I purchased three Tesla Model 3s and a Solar Panel Energy System after much strategic and financial analysis.

What happened when we became energy positive

3

empowered as our own utility? Did we make a good decision? Absolutely yes! Did we make the right financial decision? Absolutely yes! Do we have any regrets? Absolutely not!

My goal is to teach and demonstrate my strategy to you qualitatively while illustrating this quantitatively in easy-to-follow numbers.

Can anyone benefit from this strategy? It's either a Yes or No based on the following:

1) YES – You stand to benefit from this strategy if your transportation requirements demand the need for a vehicle or vehicles now and well into the foreseeable future, how much you commute in distance, if you own real estate with sufficient space to allow for full solar energy capacity coverage, and where you reside geographically.

2) NO – You will not benefit from this strategy if your only transportation option currently and in the foreseeable future is walking, biking, skateboarding, scootering, taking the bus or other forms of commuter transportation. Additionally, if you do not own real estate, you will also not be in a position to benefit from this strategy.

If the answer is yes, can you also expect to recover 100% of all of your Tesla and Solar costs? The answer is, it depends. Although some consumers will be able to recover 100% of their costs, others will not. I will expand upon this more later and explain why not everyone may be able to recover 100% of their costs.

I researched solar extensively beforehand while awaiting delivery of our Tesla Model 3s. I computed, designed, and had a properly spec'd out solar energy generation system installed. I went with solar panels.

By pairing our Tesla's with solar panels on our house, we are now fully energy independent – Energy positive, empowered as our own utility. Right now, you are saying, "That might be true, but you spent almost $190,000 to be energy independent? How can this be a winning financial strategy? No thanks! Goodbye!

I totally get it. Please read on, as my goal is to teach you why our three Tesla Model 3s and solar financial strategy, while expensive upfront, will end up allowing our family to recover all our costs, and also save our family a significant amount of money over the long run.

Decoding Elon Musk's Secret Master Plans™ was written for what the world's consumers do not yet understand and the future that is to come. My goal is that by the end of this book:

1) You will be thanking Elon Musk and Tesla for executing on his Secret Master Plans.
2) You will come to understand why Electric Vehicles and Solar are a winning financial strategy.
3) You will come to understand that the more Electric Vehicles one acquires, the more one saves due to the higher cost structure of gasoline relative to the

nominal incremental amount of solar energy needed to eliminate both gasoline and electricity costs.

4) You will gain insight into how I unlocked and decoded two key elements within Elon's Secret Master Plans.

5) You will evaluate and implement your own financial strategy by becoming energy independent: energy positive, empowered as your own utility. In doing so, you may get a free Tesla and Solar Energy System or will at least recover a good portion of your costs.

One must read Elon Musk's Secret Master Plans to understand his vision for the future. Don't worry. You can just read them. I've decoded them for you (more correctly, two key fundamentals within his Secret Master Plans).

2

ELON MUSK: THE SECRET TESLA MOTORS MASTER PLAN (JUST BETWEEN YOU AND ME) AND MASTER PLAN, PART DEUX

"The goal of Tesla is to accelerate sustainable energy, so we're going to take a step back and think about what's most likely to achieve that goal." – Elon Musk

Please take a moment to read both of Elon Musk's Master Plans. They are published online and are crucial in understanding his vision and plans for the future. I did not include the full details of the Master Plans in this book to avoid copyright infringement.

Below are the URL links as published on Tesla's website blog, recapped with highlights of each plan:

The Secret Tesla Motors Master Plan (just between you and me) – August 2, 2006: https://www.tesla.com/blog/secret-tesla-motors-master-plan-just-between-you-and-me

- Create an expensive low volume car
- Use those revenues to develop a lower priced medium volume car

- Use those revenues to create an affordable high-volume car
- Provide Solar power

Master Plan, Part Deux – July 20, 2016
https://www.tesla.com/blog/master-plan-part-deux

- Create solar roof systems integrating battery storage
- Expand Electric Vehicle products to all major segments
- Develop autonomous driving capability
- Enable your Electric Vehicle to make money for you when you aren't using it

I accidentally stumbled into decoding two key elements in these Master Plans. Yes, accidentally. Like so many others, I went straight for the highlights of Elon's plans just as referenced above. However, within both of these Secret Master Plans are details of a much, much more visionary design and strategy.

Once again, I highly encourage you to carefully read through the entirety of his Master Plans. You too might experience an epiphany discovering key elements within his plans to decode on your own. I will share with you my story of how I came to decode two key elements within Elon's Secret Master Plans.

3

THE ENERGY MATRIX™

Disclosure: Readers of this book will be introduced to new and never been used concepts, illustrations, terminologies, phrases, and principles. These are repeated numerous times throughout this book because they are new and unknown. They are needed in order to ingrain into readers' minds the fundamentals of a new energy era. Additionally, they are required in order to tie in proprietary legal intellectual property rights protection by means of copyright, branding, patent, and trademark recognitions.

"If you don't have sustainable energy, you have unsustainable energy. The fundamental value of a company like Tesla is the degree to which it accelerates the advent of sustainable energy faster than it would otherwise occur." – Elon Musk

A majority of global consumers do not realize they are living in a reality of what I refer to as **THE ENERGY MATRIX™**. Before Elon Musk and Tesla, only one energy model existed. I refer to this as the **ENERGY DEPENDENT**

MONOPOLY MODEL or MONOPOLY MODEL. The Energy Matrix dynamic came about as a direct result of Elon and Tesla implementing and executing on both of his Secret Master Plans.

The ENERGY MATRIX parallels similarities to the 1999 science fiction action film, *The Matrix*. In *The Matrix* action film, Morpheus presents Neo with two pills. Taking the red pill unveils the truth about the Matrix. Alternatively, Neo could take the blue pill, forgoing the truth, and return back to a life of blissful blind subjugation.

I will make the same analogy that a majority of global consumers have been continuously ingesting a daily black pill for most of their lives. Don't believe me? All I am offering is the truth to free your mind. All you must do is be open to it and be willing to question what is right in front of your very eyes. If you are unwilling to accept the truth, you may close this book and continue living a life of blissful blind subjugation. If, however, you are willing to unveil a new energy world order and an opportunity to free your mind, then, please continue reading to learn about this new energy paradigm created by Elon Musk and Tesla.

Excellent! You seek the truth. Let us continue. What is this daily black pill I speak of? This black pill represents consumers who purchase fossil-fueled internal combustion engine (ICE) vehicles and who, day after day, month after month, year after year, fill up and pay for their fossil-fueled vehicles, and pay their monthly electrical utility providers for their electricity consumption needs. Consumers taking this black pill are living in the tragic financial and environmental reality of what I refer to as the Energy Dependent Monopoly Model. Why a black pill? Black is the color of oil and coal. To the Oil and Electrical Utility companies, the utilization of oil and coal is black gold to them. However, for consumers, this black pill is the equivalent of black tar heroin. It's highly addictive, habit-forming, and is extremely bad for consumer's overall finances and health. Sadly, the continuous consumption of this black pill over the last century due to fossil-fuel output and consumption brought about by ICE vehicle proliferation has exacted an extremely high toll on Earth's overall health, oceans, and atmosphere.

Elon Musk and Tesla are directly responsible for ushering in a new energy paradigm encompassing electric vehicle transportation and consumer consumption of electricity, both offset by sustainable energy generation utilizing Tesla's energy ecosystem of Electric Vehicles, Solar, and Battery Energy Storage product solutions. Tesla's energy ecosystem of Electric Vehicles, Solar, and Battery Energy solutions represents a vibrant green pill to counter the realities of consumers taking the black pill. However, this green pill is more like a green jelly bean rather than a pill. Why a green jelly bean? The benefits are sweet both financially and environmentally. Vibrant green is the color of a healthy Earth tundra. A vibrant and healthy green tundra is needed to resupply the Earth's atmosphere with life-sustaining oxygen. Green is also the color of money. Green is magnificent indeed. Consumers who opt for the green jelly bean will awaken and transition into what I refer to as the **ENERGY INDEPENDENT SAVINGS MULTIPLIER MODEL™**.

This new energy paradigm changed the global energy dynamics to form The Energy Matrix and is illustrated and summarized as follows:

THE ENERGY MATRIX™

	BLACK PILL	GREEN PILL (JELLY BEAN)
	ENERGY DEPENDENT MONOPOLY MODEL	**ENERGY INDEPENDENT SAVINGS MULTIPLIER™ MODEL**
	BIFURCATED	TRINITY™
Consumer Profile:	• Drives Fossil Fuel ICE Vehicles • Does not utilize Solar Energy to power households or business electrical needs.	• Drives Electric Vehicles • Uses Solar Energy to recharge Electric Vehicles • Uses Solar Energy to power households or business electricity needs.
Benefits:	• Enriches the Oil and Electrical Utility Monopolies and Cartels at the expense of consumers finances.	• Beneficial to consumer finances. • Reduces pollution to Earth's atmosphere, oceans and soil. • Fully leverages the power of the Trinity™ model.
Side Effects:	• Is detrimental to consumer finances. • Pollutes Earth's atmosphere, oceans, and soil.	• Stops enriching Oil and Electrical Utility Monopolies and Cartels at the expense of consumers finances.

In The Energy Matrix, Tesla's new energy ecosystem created a new alternative for global consumer transportation and energy dynamics. This new transportation and energy ecosystem resulted from Elon Musk augmenting Tesla's Electric Vehicle suite by integrating Solar and Battery Energy Storage solutions in accordance with his Secret Master Plans.

Two distinct consumer energy ecosystems now exist: 1) Bifurcated (split) energy comprised of two separate energy sources: fossil fuels for powering transportation needs and electricity for powering households and businesses. The other is what I refer to as **TRINITY™** ("Energy Trinity"), comprised of Electric Vehicles, Solar, and Battery Energy Storage. Most global consumers operate under the bifurcated energy model. This model is inefficient and costly, both in terms of finances and the long-term environmental damage it causes. I will discuss the dynamics of Trinity shortly. Nevertheless, let's quickly look at sustainable energy alternatives to differentiate the energy dynamics and why Tesla's energy ecosystem is superior to them all.

Sustainable energy is typically derived and associated with four familiar sources to generate electricity: Solar, Wind, Hydro (Water), and Thermal (Heated Steam). We only need to touch on them briefly to understand their limitations and why Solar is the best solution from a global consumer standpoint.

Energy from wind is highly intermittent. It is not consistent, thereby, not a reliable source of energy. Do you have winds in excess of 30 mph consistently blowing on your property for hours, months, days, and years on end? Most likely not.

Energy from Hydro (Water) and Thermal (Heated Steam) is highly limited by geographic location. Do you have a raging waterfall or river running through your property? Do you live on top of an active volcanic molten magma chamber? Most likely not.

What about the use of hydrogen as a possible sustainable energy source? Some automakers have pursued Hydrogen Vehicles and have pushed them as the future. It's absolutely fantastic that they only emit water vapor. Unfortunately, the technology available to cost-effectively produce hydrogen on a mass scale just isn't there yet. As a result, hydrogen faces significant challenges at a consumer financial level. Ask yourself this – Can you easily and cost-effectively install a hydrogen-producing station in your garage or on your property? Most likely not.

What about Nuclear as a recurring energy source? Nuclear energy in the U.S. has been around since the first commercial central electric-generating station went live back in 1958. Can you easily and cost-effectively install a nuclear power plant in your backyard? Absolutely not.

Solar is the only sustainable energy source that is widely and cost-effectively available for a majority of global consumers to utilize. The Sun's irradiance touches almost every point on Earth. Not only is it extremely wise to use its energy, it's brilliantly (as in Sun shining) genius! The Sun's energy is totally free, widely abundant, and always returns day after day. Additionally, the Sun is expected to produce clean, free energy for at least another 5 billion years! [1] As such, this book only addresses solar energy due to its low cost, ease of adoption, and global deployment qualities coupled with the Sun's free energy for the next 5 billion years. Fossil Fuels comprised of oil and coal, on the other hand, are finite energy sources, not to mention all the

---

[1] https://solarsystem.nasa.gov/solar-system/sun/in-depth/

significant pollution that results from their utilization.

Solar energy is key to understanding Elon's vision for individuals to become energy positive by empowering themselves as their own utility. Elon was very clear on this in the first paragraph of his 2006 Secret Master Plan, *"The overarching purpose of Tesla Motors (and the reason I am funding the company) is to help expedite the move from a mine-and-burn hydrocarbon economy towards a solar electric economy, which I believe to be the primary, but not exclusive, sustainable solution."* That was back in 2006. Fast forward 16 years to present day 2022, and still, not a single other auto manufacturer, energy company, Venture Capital firm, or even Wall Street has caught on. Let us examine why Elon's vision and strategy are correct in his Secret Master Plans.

Utilizing solar for electricity generation is generally synonymous with sustainable energy. Note to the world: Sustainable Energy is so 2000. The execution of Elon's vision through Tesla led to the creation of a new energy dynamic through their highly differentiated business model. Elon Musk and Tesla upped the game in sustainable energy by creating this new energy dynamic which I refer to as Trinity stemming from Tesla's trio of Electric Vehicles, Solar, and Battery Energy Storage products. Why Trinity? Trinity refers to a group of three closely related things. Tesla's business model trio of Electric Vehicles, Solar, and Battery Energy Storage products are all highly interconnected energy products designed to work together symbiotically. Who loves this Energy Trinity ecosystem? Planet Earth, of course. Consumers around the world don't know it yet, but soon, they too will be lining up for a healthy serving of it. It's oh so delish! Trust me.

Why is sustainable energy so 2000? Let's examine the new Energy Trinity dynamic created by Elon and Tesla:

The world's Solar sustainable energy model (light green) versus Tesla's vertically integrated energy ecosystem (vibrant green) is illustrated and summarized as follows:

Everyone Else:

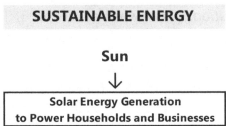

TESLA's vertically integrated energy ecosystem comprised of Electric Vehicles, Solar, and Battery Energy Storage solutions:

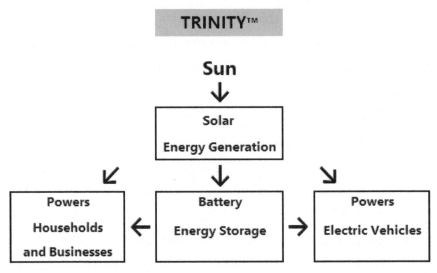

Sustainable energy (light green) is a good start for Earth.

However, Tesla is the world's only pure-play whose energy ecosystem of technology, transportation, and energy solutions symbiotically and exponentially capitalizes on solar energy utilization.   Tesla is the world's only vertically integrated company whose complementary suite of product solutions benefits consumers on a global scale, which in turn, benefits Earth the most via the Trinity framework (vibrant green).   Tesla's Energy Trinity ecosystem allows for a vibrant and healthy green Earth tundra by resupplying the Earth's atmosphere with life-sustaining oxygen by eliminating two consumer energy dependencies: Fossil-fuels and Electricity. But wait, I mentioned green is also the color of money. Don't worry.   Money will be revealed in the **ELECTRIC VEHICLE AND SOLAR SAVINGS MULTIPLIER™** model.

Let us first dive into the tragic reality of the Energy Dependent Monopoly Model within the Energy Matrix. It's wicked scary. Just how scary? It's absolutely horrifying! Don't worry. You are not alone. Most consumers have been unwillingly taking these black pills for most of their lives – a life of energy subjugation. Consumers have a never-ending, queasy feeling knowing deep down something is not right. No matter how scary or horrifying, please continue reading. Please do not close your eyes or your thoughts as we dive further into The Energy Matrix.

Remember, all I am offering is the truth. All you must do is open your mind and question that which you see and experience, day after day, month after month, year after year.

4

ENERGY DEPENDENT MONOPOLY MODEL

"We need to appeal to the people and educate them to sort of revolt against this and to fight the propaganda of the fossil fuel industry which is unrelenting and enormous."
– Elon Musk

As noted, most global consumers are living in a tragic reality subject to the Energy Dependent Monopoly Model. These consumers are fully dependent on both fossil fuel and electricity and are:

1) Consumers who drive internal combustion engines (ICE) vehicles which need fossil fuels for their transportation needs.
2) Consumers who need electricity for their households and businesses to power the many electrical devices necessary to conduct their daily livelihoods (Lights, appliances, computers, phones, etc.).

Do you believe over the long term (40+ years) that current energy prices for the U.S. for both electricity and gasoline will:

a) Decrease and cost less by the year 2060?
b) Remain unchanged and be identical by the year 2060?
c) Increase and cost more by the year 2060?

C is the correct answer.

This is further supported and evidenced by the following historical data from the U.S. Energy Information Administration for both electricity and regular gasoline costs:

U.S. Average Price of Electricity (Cents per kilowatt hour)

	1970	2021	Increase
Residential	$0.0220	$0.1419	645%
Commercial	$0.0210	$0.1176	560%

U.S. Average Price of gasoline (per gallon)

	1970	June 2022	Increase
Regular	$0.36	$4.92	1367%

Source: U.S. Energy Information Administration

Year after year, consumers are asked to conserve their fossil fuel and electricity consumption in order to reduce demand. This should translate into price reductions or at least price stabilization. Unfortunately, year after year, consumers are met with ever rising energy price increases and price volatility. Consumers need only to look at their pocketbooks and finances to know the aforementioned is true.

If you answered the question by choosing answer A or B, you might be inclined to believe that technological advancements or changing energy consumption demands in the future will result in energy price stabilization or, optimistically, lower energy prices. Unfortunately, this is not true. This is evidenced by the following:

1) Lightbulbs, electrical appliances, power tools, and equipment have been manufactured to be more energy-efficient than the prior technologies they replaced over the last century. Unfortunately, consumers continue to pay more and more in electricity costs to Electrical Utility companies because of their monopolistic, for-profit business models, and inefficient nature. LED lights are at least 80%+ more energy efficient than the incandescent bulbs they replaced over two decades ago. A 16-watt LED bulb produces the same lumen output as a 100-watt incandescent bulb. There has been mass consumer adoption to install LED bulbs as consumer's were led to believe that they would realize at least an 80% cost savings due to the lower amount of energy these so-called energy-efficient measures equated to. Refrigerators, dishwashers, washers and dryers, and other appliances are at least 30% more efficient than the ones they replaced a generation ago. Have consumers seen lower electricity rates and lower bills due to these technological advancements and reduction in consumption? Absolutely not.

2) ICE automobiles have been engineered over the last 100 years to achieve higher miles per gallon. According to the U.S. Energy Information Administration, the average fuel economy for ICE vehicles in 1970 was 12.0 miles per gallon. Gasoline has been remanufactured to be cleaner and more efficient to allow for squeezing more mileage out of every drop of gas. Since 1970, ICE engine technologies have increased nominally leading to better fuel efficiencies. According to the U.S. Environmental Protection Agency, the average fuel economy for ICE vehicles in 2020 was 25.4 miles per gallon. Have consumers seen lower prices at the pump due to these technological advancements and fuel efficiency gains? Absolutely not.

3) Over a century of multilevel technological improvements in electricity, gasoline, and their related supply and demand in the products they serve has never demonstrated long-term energy price declines. Again, gasoline and electricity prices are significantly higher than a century ago then from where they are today. There is no reason not to believe they will continue to go higher despite advances in technology, efficiencies gained, or changes in supply or demand.

4) Here is more evidence of changes in supply and demand not having an impact on lowering energy prices. California is a leader in U.S. solar installations. As of December 2021, over 1 million+ homeowner's and businesses, representing 7% of the States utility customers, have installed solar. These homeowners are mostly energy independent, drawing less energy

from the electrical grid. This should translate into lower energy prices for all consumers due to lower overall consumption and energy generation needed. However, quite the contrary. Consumers continue to face higher electricity rates despite lower energy demands resulting from the growth of solar. Less energy demand yet higher prices. Does this make sense? Get the picture? Lower electricity consumption does not translate into lower energy prices. We will explore more of this in the following chapters.

There is a growing chorus of Electric Vehicle doubters that believe fossil fuels will actually decrease in price as demand for Electric Vehicles increases. This increase in demand will result in higher electricity costs, thereby negating any benefit in transitioning to an Electric Vehicle. This argument is only half correct. It is absolutely correct that electricity prices will rise in correlation to increased consumption demand by Electric Vehicles.

Electrical Utility providers and electricity wholesalers are simply not building the electrical generation capacity and infrastructure needed to handle future exponential electricity demand. As a result, increased electrical demand through Electric Vehicle adoption could further destabilize the electrical grid causing even more outages and disruptions in the future. Electricity Utility providers and electricity wholesalers will seek to capitalize on both the increase in demand and disruptions in order to maximize profits by charging higher electricity rates.

It is also absolutely true that demand for fossil fuels will decrease in relation to ICE vehicles transitioning to Electric Vehicles. However, this will result in only a temporary decrease in prices. Consumers need to understand that oil companies are in the business of making money, not losing it. A drop in consumption leads to a direct decline in sales and revenue. In order to make up for this lost revenue due to reduced consumption, oil companies will actually increase prices over time. They cannot sustain a business model wherein their operating costs are fixed while their revenues decline. They must implement the following strategies to ensure profitability:

1) Increase prices to offset lower revenue, sales, and consumption.
2) Reduce their operating fixed costs by reducing personnel, selling off assets, not investing in new oil developments, and trimming elsewhere in general and administrative costs.

Energy prices for both fossil fuels and electricity have increased continuously for the last century, not only across the U.S. but also across the world. Consumers are continuously met with higher and higher energy bills no matter how much they conserve or reduce their consumption.

The combination of both fossil-fuel and electricity expenses represents a significant cash outflow of a consumers finances month after month and will undoubtedly increase year after year.

Consumers that maintain a strategy of purchasing or driving fossil-fuel ICE vehicles and paying for their electricity engage in a money losing proposition. This financial strategy of paying for gas and paying for electricity is money which will always leave consumer's pocketbooks, negatively impacting their finances. This strategy generously enriches the Oil and Electrical Utility companies at the expense of consumer's finances.

We will examine the power and control of Oil and Electricity Utility companies in the next chapter.

5

FOSSIL FUEL MONOPOLIES AND CARTELS

"We know we'll run out of dead dinosaurs to mine for fuel & have to use sustainable energy eventually, so why not go renewable now & avoid increasing risk of climate catastrophe? Betting that science is wrong & oil companies are right is the dumbest experiment in history by far." – Elon Musk

Fossil fuels and electrical energy are global commodities which are primarily controlled by Monopolies and Cartels. What is a Monopoly and a Cartel?

MONOPOLY – *A Monopoly is a commodity exclusively controlled by one or a few parties through exclusive ownership who dictate prices through production control and supply of that commodity.*

CARTEL - *A Cartel is a group of collaborating participants who collude with each other in order to maximize profits by controlling and dominating pricing and supply in consumer markets.*

In 1965, certain oil-rich countries banded together to form

a global fossil-fuel alliance cartel, the Organization of the Petroleum Exporting Countries (OPEC).

As of 2018, OPEC Cartel members were comprised of the following countries:

OPEC SHARE OF WORLD CRUDE OIL RESERVES AS OF 2018

Country	Crude Oil Billions of Barrels	OPEC Share	Middle East OPEC total
1 Venezuela	302.8	25.5%	
2 Saudi Arabia	267.0	22.4%	22.4%
3 Iran	155.6	13.1%	13.1%
4 Iraq	145.0	12.2%	12.2%
5 Kuwait	101.5	8.5%	8.5%
6 UAE	97.8	8.2%	8.2%
7 Libya	48.4	4.1%	
8 Nigeria	37.0	3.1%	
9 Algeria	12.2	1.0%	
10 Ecuador	8.3	0.7%	
11 Angola	8.2	0.7%	
12 Congo	3.0	0.3%	
13 Gabon	2.0	0.2%	
14 Equatorial Guinea	1.1	0.1%	
	1,189.8	100.0%	64.4%

Source: OPEC Statistical Bulletin 2019

According to OPEC's 2019 Statistic Bulletin[2], 79.4% of the world's proven oil reserves are located in OPEC Member Countries, with the bulk of OPEC oil reserves in the Middle East, amounting to 64.5% of the OPEC total.

---

[2] https://www.opec.org/opec_web/en/data_graphs/330.htm

OPEC share of world crude oil reserves, 2018

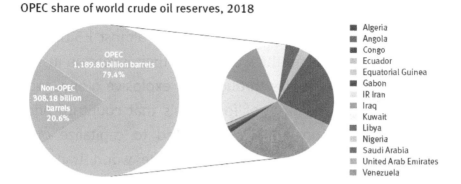

In 2016, ten additional countries were added to the OPEC cartel to form OPEC+. These countries are Azerbaijan, Bahrain, Brunei, Kazakhstan, Malaysia, Mexico, Oman, Russia, South Sudan, and Sudan.

Make no mistake. The OPEC+ Cartel was formed to exert even more monopolistic control over global fossil fuels production supply and pricing. OPEC+ now directly controls well over 80% of the world's proven oil reserves. Therefore, every consumer in the world is subjugated to whatever prices and production OPEC+ dictate.

Other top global fossil-fuel corporations control the worlds remaining reserves: Exxon Mobil, Royal Dutch Shell, Chevron, Valero, BP, Phillips 66, Marathon Petroleum, Conoco Phillips, China National Petroleum, Sinopec, PetroChina, Lukoil, Total S.A., Petrobras, Eni, and Repsol.

OPEC+ and top global fossil-fuel corporations act together in concert dictating global oil supplies simply by increasing or decreasing production output. As a result, these Monopolies and Cartels exert complete control over the world's oil supply, controlling production output and dictating prices on a massive global scale.

ICE automobiles experienced a global explosion in growth throughout the 1900s. Today, there are almost 1.5 billion ICE automobiles around the world since their invention just over a century ago. It was through this explosive global growth fueled by the demand of fossil fuels that countries and oil companies around the world rushed to capitalize on the world's transportation revolution. The world transitioned from mining gold during the Gold Rush era in the 1800s to drilling for oil, ushering in a Black Gold Rush era in the 1900s.

How significant has this black gold become? During this same time period from which ICE vehicles began production, total oil consumption went from a nominal amount to a global consumption level of almost 97 million barrels a day in 2021. One barrel of oil is equal to 42 U.S. gallons. This equates to 1.5 trillion gallons of oil consumption per year. Monopolies and Cartels were formed to capitalize on global consumer dependency for this energy commodity. These Monopolies and Cartels exert powerful control over a consumer's finances being able to control production/supply and prices. As of June 2022, crude oil prices were $125 USD a barrel. This translates into over $4.4 trillion USD equivalent in annual expense for global consumers that are lining the pockets of Monopolies and Cartels.

Consumers have few, and in most cases, no other alternatives to go elsewhere for these energy sources. Consumers must pay whatever these energy Monopolies and Cartels command. As a result, they are subject to the Energy Dependent Monopoly Model. They are dependent on both Oil and Electrical Utilities for their fossil-fueled transportation and electrical consumption needs.

I believe most consumers would agree that fuel and electricity energy prices are both volatile and will generally increase over time regardless of new technology advancements made or changes in consumption dynamics. Based on the aforementioned, consumers will be subjugated to ever increasing fossil-fuel and electricity prices at the detriment of their finances; however, much to the delight and enrichment of these energy Monopolies and Cartels.

6

GAS IN THE U.S. IS VERY CHEAP (FOR NOW)

"The reality is gas prices should be much more expensive than they are because we're not incorporating the true damage to the environment and the hidden costs of mining oil and transporting it to the U.S. Whenever you have an unpriced externality, you have a bit of a market failure, to the degree that externality remains unpriced." – Elon Musk

Believe it or not, gas in the U.S. is very cheap (for now). Gas prices in the U.S. are the envy and pale in comparison to other developed countries around the world. As of June 2022, the average price for regular unleaded gasoline in the United States was $4.92 a gallon. Most U.S. consumers are unaware of just how expensive gas prices are around the world in other developed countries. The average price for regular unleaded gasoline (in U.S. gallons and dollar equivalent) as of June 2022, were as follows:

Country	Gasoline Price Per U.S. Gallon Equivalent
Japan	$4.69
United States	$4.92
Australia	$5.36
China	$5.50
South Korea	$6.02
Canada	$6.75
Germany	$7.89
Italy	$7.99
Israel	$8.30
United Kingdom	$8.33
Spain	$8.40
Switzerland	$8.42
France	$8.57
Sweden	$9.33
Norway	$10.82
Hong Kong	$11.21

These prices are the future that awaits those consumers who continue with the strategy of using fossil-fueled ICE vehicles for transportation. Take your current fossil-fuel consumption times any of these developed countries' prices to gain perspective as to what awaits the future of your finances. These prices are reflective of June 2022, a point in time. Imagine what fossil-fuel prices will be like in 20, 25, 30, 35, or 40+ years from now? It's hard to envision, however, over a century of empirical data qualifies they will be going much higher than from where they stand from today.

Yes, it's scary, but only one-quarter scary. Why only one-quarter? Because another one-quarter scare pertains to the future of electricity, which also resembles gasoline.

What about the remaining half? It will be revealed in the following few chapters. Think of it as Stephen King writing a horror story about your future finances, only it's much, much worse. It's much, much worse because it's actually real and not just fiction. It completes the terrifying picture and full-blown horror that awaits those consumers who opt to remain in the Energy Dependent Monopoly Model. Let us next examine electricity.

7

ELECTRICAL UTILITY MONOPOLIES

"It's not as though we can keep burning coal in our power plants. Coal is a finite resource, too. We must find alternatives, and it's a better idea to find alternatives sooner than wait until we run out of coal." – Elon Musk

The invention of the incandescent light bulb in the 1870s led to lighting becoming one of the first publicly available applications of electrical power. As a result, many public utilities were created across the world as the global demand for electrical lighting and electric power consumption grew. By 2021, in the U.S. alone, there were over 700+ Electrical Utility Companies. The big player names are well known. Duke Energy, PG&E, Southern California Edison, San Diego Gas and Electric, Florida Light and Power, Consolidated Edison, Georgia Power, etc. These few handfuls of utilities control the electricity and pricing for tens of millions of consumers.

What isn't well known to consumers is that many Electrical Utilities are no longer the primary producers of electricity. How can Electricity Utilities not produce electricity? Over the last few decades, through deregulation and poor policy decisions, Electrical Utility providers, many of which are investor-owned, divested away from energy production and expansion to focus on maintaining profitability. What resulted was the creation and formation of regional electrical energy producer wholesalers by which Electrical Utilities are now dependent on for producing electricity. These Electrical Utilities made the decision to relinquish control of electricity production through deregulation to maintain profitability as they are allowed to pass these generation costs on to consumers. These Electrical Utilities are known as restructured or deregulated wholesale markets wherein they are only responsible for delivering electricity to their retail customers; electricity is generated by other entities. As a result, these generating entities monopolize and control the pricing of electricity.

Let us look at an example. Southern California Edison (SCE) serves approximately 15 million consumers over a 50,000 square mile territory. However, SCE generates less than 20% of the power it sells according to a fact sheet dated 6/14/21 from Southern California Edison's newsroom website titled – HOW SOUTHERN CALIFORNIA EDISON MAKES MONEY[3].

Here is the illustration from their fact sheet:

---

[3]

https://newsroom.edison.com/internal_redirect/cms.ipressroom.com.s3.amaz onaws.com/166/files/20179/How Edison Makes Money.pdf

HOW SOUTHERN CALIFORNIA EDISON MAKES MONEY

Updated: 6/14/21

Southern California Edison doesn't follow the typical business model of making a profit from the sale of its products and services. That is because the state of California, through the California Public Utilities Commission, wants to encourage customers to conserve power and to ensure investor-owned utilities like SCE continue investing in the electrical system infrastructure to keep it reliable. This separation of sales and profit means SCE doesn't make more money when sales increase.

WHERE YOUR MONEY GOES

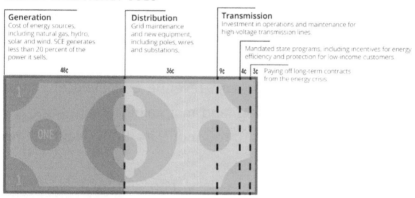

Generation
Cost of energy sources, including natural gas, hydro, solar and wind. SCE generates less than 20 percent of the power it sells.
48¢

Distribution
Grid maintenance and new equipment, including poles, wires and substations.
36¢

Transmission
Investment in operations and maintenance for high-voltage transmission lines.
9¢

Mandated state programs, including incentives for energy efficiency and protection for low-income customers.
4¢

3¢ Paying off long-term contracts from the energy crisis

*This breakdown shows SCE's costs to serve customers and implement state policies. It is based on calculations from October 2018.

According to SCE's fact sheet, "Typically, about 48 cents of each dollar pays for the cost of generating energy. Of that, nearly 39 cents is for power SCE purchases from other parties, such as developers of natural gas-fired plants and large-scale sources of renewable energy. These amounts are passed on to customers "at cost," with no markup. The other 9 cents pays for generation from facilities which SCE invests in or owns, like the Big Creek Hydroelectric Project and the Mountainview natural gas-fired plant.

About 45 cents of each dollar goes to costs of transmission and distribution infrastructure, including day-to- day operations and maintenance expenses. The remaining 7 cents includes costs for several state programs, including energy efficiency incentives.

The Commission reviews all these costs for each investor-owned utility every three years through a transparent process called the General Rate Case, which includes public input and participation by groups such as the Office of Ratepayer Advocates and The Utility Reform Network.

Through this process, the Commission allows SCE to earn a fixed return on its capital investments — the infrastructure you see every day in your communities, such as poles, wires, substations, power plants and more.

In sum, SCE does not earn revenue by selling electricity, but by building, operating, and maintaining the electric grid which serves customers throughout its 50,000-square-mile service area."

Since SCE only generates less than 20% of the power it sells, the remaining generation comes from wholesale electrical energy producers. This means SCE has absolutely no control over 80+% of the electricity produced. Consumers are thereby required to pay electrical wholesalers whatever pass-through prices they command.

Not only are these electrical generation charges passed on to consumers, so too are the delivery/distribution and operating charges by their primary Electricity provider. Delivery and distribution charges comprise fixed costs to the electrical grid in the form of wires, poles, and substations. In many cases, the electrical grid is further outsourced to electrical sub-contractors who command their own premium rates. These costs are also passed along to consumers in their electrical bills. Operating charges represent electrical utility personnel, premises, and other general and administrative costs. These costs will generally always increase and are baked into formulating ever rising rate increases to ensure these investor-owned utilities maintain their profitability goals.

Most consumers are unaware and believe that electricity is clean energy. Unfortunately, this is not the case. In 2021, according to the U.S. Energy Information Administration[4], about 4.1 billion kilowatt-hours of electricity was generated at utility-scale generation facilities in the United States comprised of the following energy sources:

---

[4] https://www.eia.gov/tools/faqs/faq.php?id=427&t=3

U.S. utility-scale electricity generation by source, amount, and share of total in 2021

	Billion kWh	% of Total
Fossil Fuels:		
Natural Gas	1,575	38.3%
Coal	899	21.8%
Other	30	0.7%
Nuclear	778	18.9%
Renewables:		
Wind	380	9.2%
Hydropower	260	6.3%
Solar	115	2.8%
Biomass	55	1.3%
Geothermal	16	0.4%
Other	8	0.3%
Total	4,116	100.0%

Source: U.S. Energy Information Administration

To recap, 60.8% of electricity was generated utilizing fossil-fuels (38.3% natural gas, 21.8% coal, and 0.7% other). This was followed by 18.9% from nuclear energy, 9.2% from Wind, and 6.3% from Hydropower. Sadly, solar energy generation only comprised a meager 2.8% of electrical energy generation. Talk about wasting free energy! We will come to see why Elon was right all along on solar energy.

Consumers might also believe that burning natural gas is clean. However, natural gas is mainly comprised of methane, a strong greenhouse gas[5]. About 117 pounds of carbon dioxide are produced per million British thermal units (MMBtu) equivalent from natural gas. So, electricity in the U.S. is less dirty (comprised of 38% natural gas and 22% coal); however, still a better alternative to gasoline.

Solar on the other hand is 100% clean and the Sun's energy is always 100% free!

---

[5] https://www.eia.gov/energyexplained/natural-gas/natural-gas-and-the-environment.php

8

ELECTRICITY IN THE U.S. IS VERY CHEAP (FOR NOW)

"We are going to exit the fossil fuel era. It is inevitable."
– Elon Musk

As of June 2022, the average residential price for electricity in the United States per the U.S. Energy Information Administration was $0.15 per kWh. U.S. consumers are still very lucky (at least for now) paying these prices. Electricity prices in the U.S. (with the exception of some states like California at $0.24 per kWh and Hawaii at $0.41 per kWh) are lower compared to other developed countries around the world. The average price per kWh (in U.S. dollar equivalent) as of June 2022 is as follows in these countries:

Country	Electricity Price Per kWh U.S. Equivalent
Israel	$0.14
United States	$0.15
Sweden	$0.19
France	$0.21
Australia	$0.23
Spain	$0.23
Switzerland	$0.23
Japan	$0.26
Italy	$0.26
United Kingdom	$0.28
Germany	$0.36

As with the fossil-fuel example, these prices are the future that awaits those who continue with the strategy of using and relying on Electrical Utilities electricity. Take your electricity consumption times any of these developed countries' prices to gain perspective as to what awaits the future of your finances. Like the fossil-fuels example, imagine what electricity prices will be like in 20, 25, 30, 35, or 40+ years from now? Electricity rates viewed over a century from empirical data qualifies they will be going much higher than from where they stand from today.

This represents the other one-quarter scare. What about the other 50% scare? Don't worry. It's coming. There are some individuals who really like horror stories but trust me on this one. It's one nobody is going to like.

9

ENRICHING BIG OIL AND ELECTRICAL UTILITY MONOPOLIES

"We are the first species capable of self-annihilation."
– Elon Musk

Remember, a Monopoly is a commodity exclusively controlled by one or a few parties through exclusive ownership who dictate prices through production control and supply of that commodity. A Cartel is a group of collaborating participants who collude with each other in order to maximize profits by controlling and dominating pricing and supply in consumer markets. These are for-profit organizations formed for the sole goal of enriching themselves and their shareholders.

Unfortunately, we cannot examine in detail all the oil profits of OPEC+ countries, but we do know they collectively earn trillions in revenue each year from global consumers. Most consumers know that Apple, Google, and Microsoft rank at the top as some of the most profitable corporations in the world. What most consumers aren't aware of is that Saudi Arabia's Saudi Aramco, of which 94% is state-owned, ranks right up there. Aramco is one of the most profitable companies in the world with a current market capitalization (as of May 2022) of $2.4 trillion. For the year ended 2021, it recorded revenues of $622 billion and posted net earnings of $200 billion. For comparison purposes, for the year ended 2021, Apple recorded revenues of $366 billion and posted net earnings of $95 billion. For the year ended 2021, Google recorded revenues of $258 billion and posted net earnings of $76 billion. For the year ended 2021, Microsoft recorded revenues of $168 billion and posted net earnings of $61 billion. The financial numbers provide perspective into just how lucrative Saudi Arabia's position is in influencing global oil production, supply, and pricing in order to drive profitability.

Let us examine earnings and executive compensation in detail of a few publicly traded Oil and Electrical Utilities in the United States based on their filings with the U.S. Securities and Exchange Commission.

For Oil Companies we will review:

1) Exxon Mobil (NYSE: XOM)
2) Chevron (NYSE: CVX)

For Electrical Utilities we will review:

1) PG&E – Pacific Gas and Electric (NYSE: PCG)
2) Edison International - Southern California Edison (NYSE: EIX)
3) Duke Energy (NYSE: DUK)

All the forgoing information is publicly available information from each company's U.S. Securities and Exchange Commission filings:

Exxon Mobile (XOM)

source: SEC Proxy & 10-K filing

Name and Principal Position	Year	Total Compensation	Year over Year (YoY) % Change
D.W. Woods	2019	$ 23,494,929	25.1%
Chairman & CEO	2018	18,777,787	7.5%
	2017	17,466,133	-
A.P. Swinger	2019	12,332,840	2.2%
Senior Vice President, PFO	2018	12,068,878	7.6%
	2017	11,215,974	-
N.A. Chapman	2019	13,304,900	38.5%
Senior Vice President	2018	9,603,006	-
	2017	-	-
J.P. Williams Jr.	2019	12,044,265	28.6%
Senior Vice President	2018	9,362,786	-
	2017	-	-
N.W. Duffin	2019	8,164,348	4.8%
President Exxon Mobil Global	2018	7,788,542	-1.2%
Projects	2017	7,879,749	-
Total		**$ 163,504,137**	
Net Income	2019	$ 14,340,000,000	-31.2%
	2018	20,840,000,000	5.7%
	2017	19,710,000,000	-
Total		**$ 54,890,000,000**	

Chevron (CVX)

source: SEC Proxy & 10-K filing

Name and Principal Position	Year	Total Compensation	Year over Year (YoY) % Change
M.K. Wirth	2019	$ 33,070,662	60.2%
Chairman & CEO	2018	20,640,623	76.9%
	2017	11,669,681	-
P.R. Berber	2019	11,420,420	42.2%
Vice President and	2018	8,030,044	-
Chief Financial Officer	2017	-	-
P.E. Yarrington	2019	9,173,170	28.1%
Former Vice President and	2018	7,159,079	-12.2%
Chief Financial Officer	2017	8,152,053	-
J.W. Johnson	2019	15,315,630	40.2%
Executive Vice President	2018	10,925,982	-0.9%
Upstream	2017	11,024,515	-
J.C. Gaegea	2019	13,009,803	65.7%
Executive Vice President,	2018	7,849,411	-12.8%
Technology, Projects & Services	2017	9,004,322	-
M.A. Nelson	2019	9,821,533	-
Executive Vice President,	2018	-	-
Downstream & Chemicals	2017	-	-
Total		$ **186,266,928**	
Net Income	2019	$ 2,924,000,000	-80.3%
	2018	14,824,000,000	61.2%
	2017	9,195,000,000	-
Total		$ **26,943,000,000**	

PG&E - Pacific Gas & Electric (PCG)

source: SEC Proxy & 10-K filing

Name and Principal Position	Year	Total Compensation	Year over Year (YoY) % Change
William L. Smith	2020 $	6,174,215	-
Interim Chief Executive Officer	2019	-	-
and President of PG&E Corp	2018	-	-
Michael A. Lewis	2020	2,082,421	198.7%
Interim President	2019	697,243	-
PG&E Company	2018	-	-
Christopher A. Foster	2020	1,080,353	-
Vice President and Interim Chief	2019	-	-
Financial Officer, PG&E Corp	2018	-	-
David S. Thomason	2020	1,527,251	131.6%
VP, CFO & Controller	2019	659,359	-7.0%
PG&E Company	2018	708,658	
John R. Simon	2020	5,128,845	231.4%
EVP General Counsel & Chief	2019	1,547,499	-50.5%
Ethics & Compliance Officer	2018	3,124,628	
PG&E Corp			
James M. Welsch	2020	2,359,190	120.3%
SVP Generation & Chief	2019	1,070,749	-
Nuclear Officer, PG&E Company	2018	-	-
William D. Johnson	2020	1,755,742	-90.5%
Former CEO & President	2019	18,529,842	-
PG&E Corp	2018	-	-
Andrew M. Vesey	2020	3,630,569	52.9%
Former CEO & President	2019	2,373,771	-
PG&E Company	2018	-	-
Jason P. Wells	2020	3,934,259	223.6%
Former EVP and CFO	2019	1,215,952	-62.0%
PG&E Corp	2018	3,197,274	-
Total		**$ 60,797,820**	
Net Income (loss)	2020 $	(1,318,000,000)	-82.8%
	2019	(7,656,000,000)	111.8%
	2018	(6,851,000,000)	-
Total		**$ (15,825,000,000)**	

Duke Energy (DUK)

source: SEC Proxy & 10-K filing

Name and Principal Position	Year	Total Compensation	Year over Year (YoY) % Change
Lynn J. Good	2020	$ 14,544,398	-3.2%
Chair President & CEO	2019	15,029,436	7.5%
	2018	13,982,960	-
Steven K. Young	2020	3,901,609	3.2%
EVP & CFO	2019	3,779,999	20.7%
	2018	3,132,755	-
Dhiaa M. Jamil	2020	4,587,700	-1.5%
EVP and Chief Operations Officer	2019	4,658,314	16.6%
	2018	3,994,400	-
Julia S. Janson	2020	4,048,848	2.4%
EVP, External Affairs & President	2019	3,955,357	47.1%
Carolinas region	2018	2,689,676	-
Douglas F Esamann	2020	3,780,353	4.8%
EVP, Energy Solutions & President	2019	3,605,920	-
Midwest/Florida regions &	2018	-	-
Natural Gas Business			
Melissa H. Anderson	2020	3,545,689	-
Former EVP, Chief Human	2019	-	-
Resources Officer	2018	-	-
Total		$ **89,237,414**	

Net Income (loss)	Year		
	2019	$ 1,270,000,000	-65.7%
	2018	3,707,000,000	-139.0%
	2017	2,666,000,000	-
Total		$ **7,643,000,000**	

Edison Int'l (EIX) - Southern California Edison

source: SEC Proxy & 10-K filing

Name and Principal Position	Year	Total Compensation	Year over Year (YoY) % Change
Pedro J. Pizarro	2020	$ 15,785,999	34.2%
EIX President & CEO	2019	11,761,702	20.3%
	2018	9,777,523	-
Maria Rigatti	2020	4,702,231	25.2%
EIX EVP and CFO	2019	3,755,630	26.5%
	2018	2,968,667	-
Kevin M. Payne	2020	4,751,027	30.8%
SCE CEO; Also SCE	2019	3,633,324	17.7%
President effective 6/7/2019	2018	3,088,108	-
Adam S. Umanoff	2020	2,775,419	6.4%
EIX EVP and	2019	2,609,613	30.3%
General Counsel	2018	2,002,170	-
J. Andrew Murphy	2020	1,848,096	-4.7%
EIX SVP	2019	1,939,857	17.4%
	2018	1,652,827	-
Total		**$ 73,052,193**	
Net Income (loss)	2019	$ 739,000,000	-42.4%
	2018	1,284,000,000	303.5%
	2017	(423,000,000)	-
Total		**$ 1,600,000,000**	

Are these Companies and handful of Executives entitled to earn this compensation and profit from it? Absolutely! They should not be reviled for their profit efforts. They are capitalizing on their efforts in fulfilling global consumer energy needs. These companies have collectively spent hundreds of billions over many decades investing and building up the infrastructures and supply chains needed to satisfy consumer energy demands.

Now, are there instances of abuses and negligence sprinkled into energy pricing every now and then? Sure, there are. I won't go into multiple examples as one only needs to do their own research on the internet to read the many, many stories of energy pricing abuses.

Here are just two small recent examples of such abuses. On June 7, 2022, the U.S. National average for a regular gallon of gasoline reached a record high of $4.92. In California, gas also reached a record high of $6.37 for a regular gallon of gasoline. Despite the record high prices, look no further than California compared to the U.S. National average for examples of price gouging making national news headlines. Yes, California's gasoline is more expensive than the U.S. National average due to added taxes and fees. About $1.26 in taxes and fees are baked into gasoline costs. These taxes and fees are as follows:

Taxes:	
Federal Excise Tax	$ 0.18
State Excise Tax	0.51
Sales Tax (estimated)	0.18
Fees:	
Low Carbon Gas Programs	0.17
Greenhouse Gas Programs	0.20
Underground Tank Storage	0.02
Total fees and taxes	**$ 1.26**

Evidence of price gouging was recently apparent as news articles and media reports dated June 3, 2022, noted gas prices at one Chevron, in Los Angeles, California topped $8. It was even worse at a Chevron in Mendocino, California that approached nearly $10 that same day. If the U.S. National average was close to $4.92 and you add California's fees and taxes of $1.26 that puts gasoline in the range of $6.18 a gallon. How these two Chevron's managed to charge above $8 and near $10 is representative of egregious price gouging.

Chevron gas station in Los Angeles, California on June 3, 2022

Chevron gas station in Mendocino, California on June 3, 2022

These were just two examples of gasoline gouging. How about electricity? A 2019 audit investigation severely criticized The Los Angeles Department of Water and Power (LADWP) for allowing excessive overtime. In 2018, 306 of LADWP's workers took home more than $100,000 in overtime pay, while the agency paid a total of $250 million for overtime. A shocking example of this excess overtime was a security worker with a base salary of $25,000 who was paid $314,000 in overtime. He was not alone. There were three additional peers who were paid more than $200,000 each in overtime. (The nationwide median wage for security officers was $28,500 in 2018.) What's even more shocking is that these payouts were all legal under terms spelled out in their union contract. The unions contract is designed to enrich workers to the greatest extent possible enabling hefty overtime payouts. A provision in the unions contract requires a normal shift worked after more than one hour of overtime to be paid at double time, as well as that overtime is not based on working more than 40 hours in a week, but on working time beyond a "normal" shift. A separate study found that LADWP's yearly payroll expense per customer was $490, significantly higher than the nationwide median for large utilities of $280 per customer. These gross excesses are simply passed on in the form of higher rates translating into higher consumer electricity bills.

The Oil and Electricity industries each enjoy complete control and pricing power over fuel and electricity costs. Consumers and businesses who do not have an Electric Vehicle and a Solar Energy generation system must pay whatever rates they demand to refuel their gasoline vehicles and supply electrical power to their homes and businesses.

You may think the aforementioned wasn't scary at all. You may just be saying this was more of an enlightenment of the future of energy prices and the enrichment the energy industry stands to make from those consumers who continue with the strategy of buying/driving ICE vehicles and paying their monthly electricity bills. The real scare comes when I quantify financially just how tragic and horrifying this ongoing strategy will be on consumers' future finances.

Are you ready to take a vibrant green jelly bean yet? I assure you, it's totally sweet.

Be prepared to be awakened in The Energy Matrix as we dive into the Energy Independent Savings Multiplier Model or Multiplier Model.

10

ENERGY INDEPENDENT SAVINGS
MULTIPLIER™ MODEL

"Obviously, Tesla is about helping solve the consumption of energy in a sustainable manner, but you need the production of energy in a sustainable manner." – Elon Musk

Using the Sun's free energy for solar energy generation is a natural hedge to Electric Vehicles and households or business's electricity needs both financially and environmentally. As such, Electric Vehicles that are paired with and recharged by Solar Energy engage in a complementary symbiotic financial and environmental hedging strategy that allows for consumers to independently power both their transportation needs and their homes or business electrical needs. In doing so, they eliminate their fossil-fuel and electricity expense dependencies while simultaneously eliminating their carbon emission output.

Solar can be installed on almost any roof. Solar can also be ground mounted and can also be raised above ground, such as a Solar Carport. Premium solar panels typically come with a 25 year or higher warranty. However, premium solar panels are capable of producing energy well beyond their warranty coverage periods. Our solar panels are warrantied for 25 years. However, they are designed to produce energy for up to 40 years. Just like how most consumer ICE vehicles are given a general 3-to-5-year warranty, most ICE vehicles are designed to last well in excess of the warranty when the warranty period ends. As of 2020, the average U.S. vehicle age was 12 years old, further illustrating the point that products do last much longer than their stated warranties. The same warranty and energy production life attributes also apply to Solar.

A Solar Energy generating system in the form of solar panels does many important things for our family:

1) Solar panels provide all the energy needed to recharge our three Tesla Model 3s. As a result, we have our own recharging power station. This is the equivalent of having our own gas station.

2) Our solar panels also provide all the energy needed to also power our household electrical consumption needs. The equivalent of having our own electricity utility power plant.

3) We paid a one-time, fixed set cost for solar in exchange for realizing up to 40 years of future energy production to cover both our transportation and household electricity consumption needs. As a result, we are no longer subject to energy price fluctuations and most importantly, we are no longer subject to future energy price increases.

4) Our solar panels are roof mounted. This extends the useful life of our roof and helps to keep the inside of our property cooler during the hot summer months. This further aids in reducing the need to operate our air conditioning system.

Sounds great, doesn't it? Imagine owning a Toyota, Honda, Ford, or VW and having your own gas station with a limitless supply of gasoline. Now couple that with having your own electrical power station like Duke Energy, Southern California Edison, PG&E, etc. with an infinite supply of electricity. It would be great if these energy sources were free, but they aren't.

Most consumers realize these energy costs are expensive today. However, they fail to realize the magnitude of how detrimental these energy price increases will be on their finances in the future. These expenses have conditioned consumers' mindsets to believe expenses for gasoline and electricity are simply what they are. They tend to focus on the month-to-month expenses (short view) rather than the cumulative financial impact these expenses have over many, many years on their finances (long view).

Most consumers view Electric Vehicles as expensive compared to ICE vehicles. While this may be true in regard to the upfront purchasing costs, this is a short-sighted and misguided financial view. What is more expensive than a Tesla? Answer: The future price of gasoline and electricity. What I developed and demonstrate is that there is a winning financial strategy for consumers who take the long view perspective on Electric Vehicles by utilizing Solar Energy.

It is without question that consumers' finances continue to be negatively impacted as more of their discretionary incomes are directed at paying higher gasoline and higher electricity bills. It was previously qualified that both of these energy costs will only increase in the future regardless of any new advancements in technologies, efficiencies gained, or changes in supply output or consumption demand.

I spec'd out our Solar Energy System to ensure we are energy positive. We paid a fixed, one-time cost for our Solar that will cover our Tesla Model 3s transportation and our household's electricity needs for the next 40 years. By eliminating these two energy dependencies, we eliminate future financial exposure and risk to our family's finances. We are now fully energy independent and price-protected (hedged) against what consumers are witnessing now in rising gasoline and electricity prices and will continue to see well into the future.

Let me repeat that. The moment we strategically bought our three Tesla Model 3s and paired them with a properly spec'd out solar energy generation system, we instantly became financially energy independent. We are in control of our own energy needs and we significantly reduced our CO_2 emissions. We are no longer tragic financial victims of the Energy Dependent Monopoly Model. We are now protected (hedged) for the next 40 years against three important things:

1) We are price-protected against both gasoline and electricity energy price volatility (increasing and decreasing price movements).

2) We are price-protected against both gasoline and electricity energy future price increases (monopolistic greed, inflation, price gouging, and poor policy decisions) and;

3) We immediately and significantly reduced our environmental CO_2 output.

We are successfully positioned for both energy financial freedom and environmental success for the next 40 years as a result of becoming energy independent by eliminating both gasoline and electricity costs. By undertaking this financial strategy, we are no longer enriching the Monopolies and are free from the Energy Dependent Monopoly Model.

I will qualify that Electric Vehicles and Solar are not a panacea solution for every consumer or business. While the sun irradiates its energy across the Earth, different geographic regions in higher latitude settings receive lower solar irradiance and experience longer periods of prolonged darkness due to changing seasons relative to the Earth's axial tilt. Barrow, Alaska is an extreme example where such a strategy would most likely not be beneficial. During the winter months, Barrow experiences 24 hours of complete darkness for nearly 2 months. Add to this the extremely cold temperatures which would greatly dissipate the energy capacity for an Electric Vehicle coupled with high levels of snow which would impede Solar Energy production. This is just one example. Another example would be in dense Metropolises such as Manhattan, New York. Manhattan geographically is at a higher latitude with almost 1.7 million citizens living in a land area of 23 square miles coupled with significant high rises and multifamily housing. The combination of all these factors dictate that an Electric Vehicle and Solar strategy would be a challenging solution due to the limited amount of roofing space available combined with the population density and the consumers related energy needs. Contrast this with Los Angeles County, California set in a lower latitude which also boasts a high population density. Los Angeles County, however, is spread out over a larger geographic area, has few high rises, and boasts a high concentration of homes and land for solar to be deployed. Elon Musk was clear about this in the last sentence of the first paragraph in his 2006 Secret Master Plan noting, *"the overarching purpose of Tesla Motors (and the reason I am funding the company) is to help expedite the move from a mine-and-burn hydrocarbon economy towards a solar electric economy, which I believe to be the primary, but not exclusive, sustainable solution."* It would be great if a solar electric economy was a 100% fit's all solution. The reality is not every consumer or business will be able to leverage the full benefits of Trinity.

Still, most global consumers will be able to leverage an Electric Vehicle and Solar savings financial strategy. Monthly recurring energy expenses for both gasoline and electricity which are directed into the Monopoly's pocketbooks should instead be evaluated and re-directed towards an Electric Vehicle paired with a Solar Energy System that provides consumers with up to 40 years' worth of independent energy.

The aforementioned recaps the framework within the **Energy Independent Savings Multiplier**™ model (Refer to table on page 12). But how does being energy independent become a savings multiplier? When I refer to being energy independent, I refer to the combination of Electric Vehicles and Solar. As such, I built an **ELECTRIC VEHICLE AND SOLAR SAVINGS MULTIPLIER**™ ("Multiplier") financial model.

I designed the Multiplier as a straightforward model that financially unlocks what Musk conveyed to the world in his Secret Master Plans. What he conveyed to the world in words, I decoded and financially convey into numbers. The Multiplier model financially illustrates the combination of Electric Vehicles and Solar Energy as a means for consumer energy independence by becoming energy positive, empowering oneself as their own utility. The transformative, hard number results are more than illuminating, they are jaw-dropping! It completely validates Elon Musk's vision and direction for consumer energy independence by combining Electric Vehicles and Solar Energy. I will qualify and demonstrate why they are a winning financial strategy.

Creating a straightforward, easy-to-follow dynamic financial model that breaks down multidimensional financial elements of gasoline and electricity consumption, usage, and related future energy costs fully offset by Solar Energy generation has never been done before. No Solar Company or Electric Car Company (not even Tesla) has such a financial model.

Tesla's suite of products is already accelerating the world's transition to Electric Vehicles and Solar Energy. Elon and Tesla debuted the new Roadster in 2018 to give a hard-core smackdown to gasoline cars. The Multiplier financial model I designed delivers a triple fatal blow to gasoline ICE cars, the Oil industry, and Electrical Utilities. I built it for the entire United States in just under two days.

To understand why Tesla's combination of Electric Vehicles and Solar Energy are a winning financial strategy, I approached this both qualitatively and quantitatively, the way I do with any strategy. It is easy to describe qualitatively, which is the purpose of this book. I created the financial model to demonstrate it quantitatively. Qualitative addresses the words and meaning behind the quantitative numbers and data. Both are needed to support each other to provide perspective and guidance to the information being conveyed.

The qualitative part started at the very beginning of THE ENERGY MATRIX chapter. Let us continue and examine the financial aspects of why Elon's vision for consumer energy independence and why the Electric Vehicle and Solar Savings Multiplier model make financial sense.

11

ELECTRIC VEHICLE AND SOLAR SAVINGS MULTIPLIER™

"If you have a great solar roof, and you have a battery pack in your house, and you have an electric car, that scales worldwide. You can solve the whole energy equation with that." – Elon Musk

Most consumers who already utilize solar energy to provide for their household electricity consumption understand the cost savings benefit of eliminating electricity expenses. What no one in the Solar Industry or Electric Vehicle Industry has yet determined is that when Electric Vehicles are paired with Solar and Battery Energy Storage to power both transportation and electricity consumption needs, that combination becomes an **Energy Multiplier™** under the Electric Vehicle and Solar Savings Multiplier framework. As a result, these consumers directly benefit from being able to capitalize on a **Power Multiplier™**, being able to power both their Electric Vehicles and their homes or businesses. Additionally, these consumers also directly benefit from being able to capitalize on a **Charge**

Multiplier™, being able to **Charge Free and Clean**™ with Solar for their homes and businesses, and **Recharge Free and Clean**™ with Solar for their Electric Vehicles and Battery Energy Storage. This results in an **Environmental Multiplier**™ effect as a result of eliminating both gasoline and electricity dependencies. To recap, consumers that are Energy Positive can leverage an Energy Multiplier, a Power Multiplier, a Charge Multiplier and an Environmental Multiplier. In doing so, Energy Positive consumers will also realize a **Savings Multiplier**™. As discussed in the Energy Matrix, this is where green, also the color of money, comes into play. Using Solar Energy alone (light green) to realize solar savings is already a compelling value proposition. The Multiplier model illustrates both Electric Vehicles and Solar as a winning financial strategy that becomes a Savings Multiplier (vibrant green) for consumers' finances when compared to Solar Energy alone.

The differentiation in Solar Savings alone versus the Electric Vehicle and Solar Savings Multiplier is illustrated as follows:

ELECTRIC VEHICLE AND SOLAR SAVINGS MULTIPLIER™

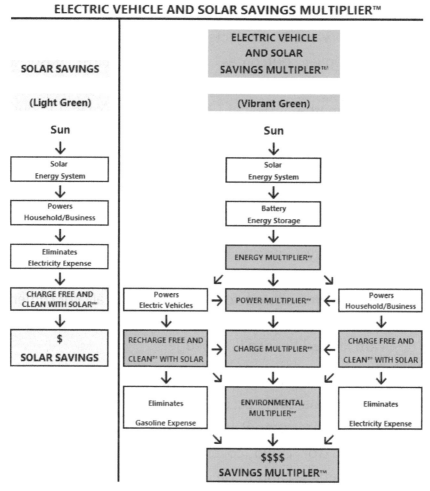

Now comes the moment you have been waiting for. What is the **Electric Vehicle and Solar Savings Multiplier™** model?

Queue the drumroll and take a deep, deep breath. No really! I'm serious now! Breatheeeeeeeeeeee and....go! The Multiplier model I created in technical terms is an online proprietary non-downloadable data model for use in connection with consumer electricity consumption, electric vehicles electricity consumption, offset with solar energy generation using energy sector metadata utilizing dynamic algorithms which model dynamic "what-if" scenarios to demonstrate various projected future financial money savings multiplier outcomes compared against gasoline vehicles, gasoline fuel charges and utility electrical costs....and breathe...... Haha! I did warn you. Yeah, I know I totally just lost 99.9% of you out there. Don't worry. I created two Multiplier models:

1) One that is a high-level model geared for Elon Musk, Tesla's C-Suite/ Board of Directors, Tesla's Finance, and Energy Teams.
2) One for everyone else. A consumer-level, step-by-step model geared for Tesla Owners, all consumers interested in Electric Vehicles, Solar Energy, and those with limited finance knowledge.

I want to make it absolutely clear however, that no Solar model can ever be 100% accurate. While the Sun rises every day, Mother Nature cannot be modeled. The frequency and time duration for cloud coverage, rain, snow, and other weather dynamics cannot be predicted with any kind of precision in the future. Additionally, Energy Monopoly's and Cartels, World Governments, and geopolitical energy policies will always act in their best interest in dictating both energy costs and supply dynamics. As such, their actions also cannot be predicted or modeled with any kind of certainty in the future.

Although Solar may seem simple, there are additional aspects in achieving being energy positive which consumers need to take into consideration as they adopt Solar in order to maximize solar energy production. I refer to them as **Solar Energy Dynamics™** comprised of the following factors:

1) **Geographic Location** – Solar energy generation and production is highly contingent on the amount of sunshine received, also known as solar irradiance/solar insolation, based on your geographic location.

2) **Cardinal Solar Panel Orientation (North, South, East, West mounting)** – In the United States, the sun rises and sets in the southern sky. As a result, solar energy generation is optimal and most efficient when solar panels are positioned south-facing.

3) **Solar Panel Mounting Tilt (Azimuth tilt)** – The angle of degree or pitch to which solar panels are installed relative to their geographic location.

4) **Solar Cell Composition** – Monocrystalline versus Polycrystalline.

5) **Solar Efficiency** – The higher the solar panel efficiency, the higher the lifetime energy production yield.

6) **Solar Degradation** – The lower the degradation, the higher the lifetime energy production yield.

7) **Solar Space Capacity** – The more space available, the more panels that can be installed to achieve being energy positive.

8) **Battery Storage** – After hours Energy from Sunset to Sunrise + Backup Energy Storage. A key element in Elon's Master Plan.

9) **Warranty Coverage** – Not all warranties are equal.

10) **Solar Company's Sales Knowledge and Experience of Installation Team** – Not all are equal.

Although I would like to expound upon them more in detail, I am reserving the aforementioned for another book as I didn't want to lose a reader in the more technical aspects of Solar. The main key to remember with being energy positive is producing more energy than one consumes.

Now back to the Multiplier. The Multiplier quantifies financially what happens when an individual becomes energy positive, empowered as their own utility as specified in Elon Musk's Master Plans.

To access the model, go to www.energytrinity.com. I recommend waiting until completing this book to access the model. (Note: The online model is currently built for any consumer to use in the United States. All I humbly ask is that if you find the information beneficial, please consider making a voluntary donation. It takes time and resources to continuously update energy metadata. I will initially try for a donation-based platform however, I may need to later charge a nominal fee to utilize the model.) The Multiplier model demonstrates the tragic future financial horrors that await consumers who continue to buy/drive ICE vehicles and pay for their electricity versus those that elect the strategy to buy/drive Electric Vehicles and power both their transportation and household or business electricity needs with Solar Energy.

As mentioned earlier, it delivers a triple fatal blow to ICE gasoline cars, the Oil industry, and Electrical Utilities. It's designed for those consumers who are looking for something big, bold, robust, and more precise in order to make a more informed financial decision. It quantifies Elon's vision of being energy positive by empowering an individual as their own utility. Does an Electric Vehicle paired with a Solar Energy System make financial sense? More importantly, what qualifies me and the Multiplier model in quantifying the validity of the data? Read the chapter on "Connecting the Dots and Pacific Coast Banking School" to see if I am worthy of such financial and modeling qualifications to earn your consideration.

Don't like financial models? Understandable as finance can be just darn right confusing to most people. I will provide a free, high-level basic option. The basic financial analytics is very straightforward. I would caution, however, that the preceding example is extremely high level basic. It does not cover the dynamic in achieving being energy positive or provide projections of future energy costs for both gasoline and electricity. Yes, being energy positive on its surface is a very straightforward concept. However, not all solar is equal. Solar is much more nuanced and where it can all go wrong if not thought out well in advance. How wrong? Most solar installations done to date are inadequate in terms of energy production, leaving open the possibility of risk exposure. What exposure? Needing to draw on the electrical grid, leaving one at risk of future electrical bills. The Multiplier model addresses much of the dynamics needed in order to achieve being energy positive. Although Solar Energy isn't new, the idea's pertaining to the financial benefits of both Electric Vehicles and Solar presented in this book are. As noted, Solar is where it can all go wrong and deserves another book in its own right in order to maximize Solar Energy production to mitigate future risk exposure.

Let's explore a few free high-level hypothetical gasoline and electricity expense examples. Interestingly, no Solar Company, Electric Vehicle Company or even Tesla has yet to qualify or quantify savings metrics related to the combination of Electric Vehicles and Solar. If they can't see it, it's even more understandable why consumers cannot see what the future holds in store for their finances. For a quick back of the envelope, high-level calculation, take your annual electricity bill and times this by 40 years, and then take your annual ICE vehicle fuel expense and times it by 40 years. For example, if your monthly average electricity bill is $175 a month, times it by 12, which equals $2,100. Now times $2,100 by 40 years. This equates to $84,000 in electricity expense based on current electricity rates and consumption (assuming these do not change).

Why a 40-year period? Remember, a premium solar panel system can provide up to 40 years of energy - much longer than the stated warranty period coverage. Does it make financial sense to pay, for example, $20,000 today for a Solar Energy System to eliminate $84,000 in future electricity expenses?

Remember, $84,000 is just a flat assumption assuming electricity rates stay the same for 40 years. If you believe electricity rates will be higher, then you can assume an even higher amount. Let's assume electricity rates are going higher over 40 years and that it is estimated that instead of $84,000, it will be $100,000 or $125,000. How do you know what future prices will be? This is where the online financial model comes into play with the proprietary data algorithm modeling. It projects the future costs of both electricity and fossil-fuel expenses.

Now getting back to the free example, does it make financial sense to pay $20,000 today for a Solar Energy System that will eliminate $100,000 or $125,000 in future electricity expenses over the next 40 years? If so, then you would most likely conclude that Solar is a compelling value proposition to eliminate both electricity volatility and pricing risk over the next 40 years. As a result, undertaking such a financial strategy might make solar a compelling and great value-added home or business improvement project.

Now let us look at gasoline expenses. If your monthly fuel expense for one vehicle is $200 a month, times that by 12, which equals $2,400. Now times $2,400 by 40 years. This equates to $96,000 in ICE fuel expense based on current fuel prices and consumption (assuming they do not change).

Now add $84,000 + $96,000. This equals $180,000 in combined electricity and ICE fuel expense. This assumes that both electricity and fuel pricing and consumption remains unchanged for the next 40 years. Ask yourself? Do you believe electricity and fuel prices will decrease, stay the same, or increase over the next 40 years? I believe most consumers would agree both gasoline and electricity will be higher based on empirical data over the last 100 years confirming such increasing trends and prices regardless of any new technologies, increased efficiencies, or changes in supply and demand.

Now, let's move on to the Multiplier model effect to understand why the combination of Electric Vehicles and a Solar Energy System is a winning financial strategy. There are still those doubters of Electric Vehicles who believe they are expensive. However, let's see if I can get them to reconsider this widely held view.

All things being equal, most consumers need vehicle transportation (there are just over 275 million registered vehicles in the U.S.). Let's say a consumer is evaluating two options: 1) Spend $30,000 on an ICE vehicle or 2) Spend $50,000 for an Electric Vehicle. The Electric Vehicle represents a $20,000 premium above consideration over an ICE vehicle.

If a consumer chooses to purchase and drive an ICE vehicle, their income will always be directed towards fuel costs which will always go to enrich the Oil Monopolies. However, what if a consumer opts to change strategies and purchase an Electric Vehicle and have it recharged by their own Solar Energy System? By adopting this strategy, those fuel and electricity expenses which would have gone to enrich the Oil and Electrical Utility Monopolies are now redirected towards their Solar investment and Electric Vehicle.

Let's go back to our example. A consumer makes the decision to go with an Electric Vehicle and install a Solar Energy System to power both their transportation and electricity needs. More Solar panels are now required due to the additional electrical energy consumption requirements. So instead of paying $20,000 to cover just their electricity needs, let's say a $25,000 investment is now needed for a Solar system that covers both their transportation and household or business electricity needs.

Does it make financial sense to pay $25,000 today for a Solar Energy System to eliminate $180,000 in future electricity and fuel expenses over the next 40 years? Remember, $180,000 is just a flat assumption assuming electricity and fuel prices remain unchanged from today. If you believe electricity and fuel prices will be higher in the future, then you can assume an even higher amount.

Let's assume electricity and fuel prices are going higher over 40 years and that it's estimated that instead of $180,000, it will be $225,000, or $250,000. (Remember, the online Multiplier quantifies projected energy costs to better understand what these costs might be in the future.) Now, does it make financial sense to pay $25,000 for a Solar Energy System today that will eliminate $180,000, $225,000, or $250,000 in future electricity and fuel expenses over the next 40 years? If so, then you would most likely conclude that the combination of an Electric Vehicle and a Solar Energy System is a winning financial strategy to eliminate both gasoline and electricity volatility and pricing risk over the next 40 years. As a result, undertaking such a financial strategy would absolutely make it a must-have value-added home improvement project and an Electric Vehicle acquisition proposition.

But what if a consumer wishes to acquire more than one Electric Vehicle and additional Solar Energy? We will also address this scenario and how the Multiplier comes into play.

(Note: The online Multiplier accounts for this and allows a user to model up to 12 vehicles.)

Let's recap and break down some of the scenarios/strategies from the preceding example and then look at the acquisition of two Electric Vehicles and additional Solar Panels:

Strategy 1 - Continue Buying/Driving ICE Vehicle and Paying for Gasoline and Electricity

This strategy is always a money losing proposition as consumers' incomes will always go towards enriching the Monopolies. (This strategy assumes gasoline and electricity prices remain unchanged over the next 40 years):

	Monthly	Annual	40 Years
Gasoline expense	$ 200	$ 2,400	$ 96,000
Electricity expense	175	2,100	84,000
Total	$ 375	$ 4,500	$ 180,000

Strategy 2 – Eliminate Electricity only:

What if a consumer can't part with their ICE vehicle but wants to eliminate their electricity expense? Using the example, if a consumer invested $20,000 today in a Solar Energy System to eliminate $84,000 in future electricity expenses over 40 years, the consumer recovers 100% of their initial $20,000 Solar Investment. This strategy would make their system free and saves them $64,000+ by eliminating 40 years of future electricity expense. Sound like a compelling value proposition? The only problem is that this still leaves a consumer's finances open to $96,000+ in future gasoline

72

expense risk exposure by keeping the ICE vehicle:

	Monthly	Annual	40 Years
Electricity Costs Eliminated	$ 175	$ 2,100	$ 84,000
Solar Energy System	-	-	(20,000)
Total Electricity Savings	$ 175	$ 2,100	$ 64,000
Gasoline Expense Risk Exposure	$ 200	$ 2,400	$ 96,000

Strategy 3 - Acquire Electric Vehicle and Solar to Eliminate Gasoline and Electricity:

This strategy is where the Multiplier methodology comes into play. It demonstrates a projected future financial savings multiplier outcome relative to the cost of the Solar investment.

A multiplier is represented as "x" in arithmetic form. A multiplier of "x" from a financial application is simply a factor that increases the base value of the item being discussed. The "x" multiplier will be discussed and shown in the next few strategy examples. Interestingly, X.com and SpaceX were also co-founded and founded, respectively, by Elon Musk. "X" is significant at Tesla. However, I will touch more on this later.

In Strategy 2, we assumed a $20,000 Solar Energy System investment today would eliminate $84,000 in future electricity expenses over the next 40 years. This translates into a 4.2x savings multiplier ($84,000 saved in exchange for $20,000 invested or $84,000 divided by $20,000). However, this strategy for solar alone still leaves the consumer open to future gasoline risk due to their decision to stay with an ICE vehicle. The $96,000 in future gasoline expense wipes out the entire solar savings benefit.

In Strategy 3, we acquire an Electric Vehicle and Solar Energy System. Additional solar energy is required to satisfy additional electrical charging needs for transportation costing $5,000 more. The example notes a $25,000 Solar Energy System investment today eliminates both gasoline and electricity expenses saving the consumer $180,000+ over the next 40 years. This translates into a 7.2x savings multiplier ($180,000 saved in exchange for $25,000 invested or $180,000 divided by $25,000).

We see that the savings not only covers the $20,000 premium the consumer paid above the ICE vehicle, but it also pays for the entire $50,000 Electric Vehicle itself! The net result is that the consumer recovers 100% of their $25,000 Solar investment and their entire $50,000 Electric Vehicle and will still save a net $105,000+ in the future over 40 years due to not having to pay for both gasoline and electricity. Their Solar Energy System and Electric Vehicle investments are 100% cost recovered! This represents a 4.2000x net savings multiplier above the $25,000 net solar investment. Both the Electric Vehicle and Solar thereby become free! Yes, Free!

A net Multiplier is shown to provide a clearer picture of future expenses avoided less the investment made towards an Electric Vehicle and Solar Energy System as illustrated below:

Net Solar Investment	$ 25,000	[A]

	40 Years	
Gasoline Expense	$ 96,000	
Electricity Expense	84,000	
	$ 180,000	[B]
Less:		
Electric Vehicle	(50,000)	
Solar	(25,000)	
Total Net Savings	105,000	[C]
Multiplier	7.2x	[B] / [A]
Multiplier, net	4.2x	[C] / [A]

If this makes financial sense to you, then you now understand the basis of the Multiplier model.

The Multiplier represents a multiple of Electric Vehicle and Solar financial money cost savings above just the solar savings option by itself through the elimination of gasoline and electricity. (Note: It can also be calculated out as a Return on Investment or ROI. I chose to use a multiplier however, due to other dynamic elements in this book that leverage off the word multiplier. Try plugging in ROI every time multiplier is used and you will understand why I used a multiplier).

This strategy is an example of spending money now to save money in the future. Still think Electric Vehicles are expensive? Or, is the future of energy over 40 years expensive? Are you having second thoughts about that ICE vehicle now? Remember, gasoline and electricity expenses will always be a drain on consumers' finances – an always money-losing strategy. A Solar Energy System paired with an Electric Vehicle becomes an investment strategy allowing a consumer to recapture those costs by eliminating gasoline and electricity expenses.

Let's take it one step further and add another Electric Vehicle and additional Solar Energy.

Strategy 4 – Acquire Two Electric Vehicles and More Solar to Eliminate Gasoline and Electricity:

Let's say a consumer is evaluating two options. We will double it for simplicity: 1) Spend $60,000 on two ICE vehicles or 2) Spend $100,000 on two Electric Vehicles. The two Electric Vehicles now represent a combined $40,000 premium above consideration over the ICE vehicles. Let us also assume that now a $30,000 investment is needed for more Solar Energy production instead of $25,000 in order to cover household or business electricity needs and additional transportation needs. Let us also assume all things are equal and just double the annual gasoline expense from $96,000 x 2 = $192,000. Electricity expense is still $84,000. Both of these expenses add up to $276,000 over 40 years. This translates into a 9.2x multiplier savings ($276,000 saved in exchange for $30,000 invested or $276,000 / $30,000).

The savings not only covers the $40,000 premium the consumer paid above the two ICE vehicles it also pays for both Electric Vehicles totaling $100,000! The net result is that the consumer recovers 100% of their $30,000 Solar investment and their entire $100,000 Electric Vehicles and will still save a net $146,000+ in the future over 40 years due to not having to pay for gasoline or electricity. Their Solar Energy System and Electric Vehicle investments are 100% cost recovered, and they will save an additional $146,000+ through the elimination of both gasoline and electricity expenses once their investments are fully cost recovered. Their two Electric Vehicles and Solar also becomes free! Yes, Free! This represents a 4.8667x net savings multiplier above the $30,000 net solar investment. A net Multiplier is shown to provide a clearer picture of future expenses avoided less the investment made towards an Electric Vehicles and Solar Energy System as illustrated:

Net Solar Investment	$ 30,000	[D]

	40 Years	
Gasoline Expense	$ 192,000	
Electricity Expense	84,000	
	$ 276,000	[E]

Less:		
Electric Vehicles	(100,000)	
Solar	(30,000)	
Total Net Savings	146,000	[F]

Multiplier	9.2x	[E] / [D]

Multiplier, Net	4.8667x	[F] / [D]

The aforementioned example demonstrates that the more Electric Vehicle's one acquires relative to the solar investment made can be an even greater savings multiplier. This is where the absolute beauty of the Multiplier reveals itself: That the more Electric Vehicles one acquires, the more one saves due to the higher cost structure of gasoline relative to the nominal incremental amount of solar energy needed to eliminate up to 40 years of both gasoline and electricity costs. It is an element I accidentally unlocked within Elon's Secret Master plans in building my Multiplier models.

The Multiplier dynamic is even more apparent today. When I built the Multiplier back in 2018, the average price for regular gasoline in the U.S. was $2.74 a gallon. As of June 7, 2022, the average price for regular gasoline in the U.S. stood at $4.92 a gallon. Now, this doesn't mean going out and buying multiple $135,000 Tesla Plaid Model Ss for you, your spouse, or your entire family. It would be fantastic if you could recover 100% of these costs too! However, I believe most consumers would agree and understand that acquiring two Tesla Model 3s or Model Ys totaling say $120,000 - $130,000 would save more than a consumer buying two Plaid Model S's totaling $270,000. I go into detail in the next chapter of how we are recovering 100% of our Tesla's and Solar Energy System. It is very straightforward. The examples walk one through our cost outlays and energy costs incurred and avoided. It's what a typical accountant would put together. Yuck, I know, no one likes accounting. This is why I needed to also build the Multiplier financial model as it allows a consumer to input various cost and energy elements to render meaningful financial outcomes to make more informed decisions.

So, you see, the Multiplier is very simple once one steps back and calculates out some basic numbers. Mind-blowing right? Well, that was crazy easy! It is said everything is hard until it is discovered, and then from there, it's simple. It really is that simple. So simple, that no one has figured it out – but I did. ;-)

Here is a summary illustration of the strategies and Multiplier effects from the preceding example:

MULTIPLIER™

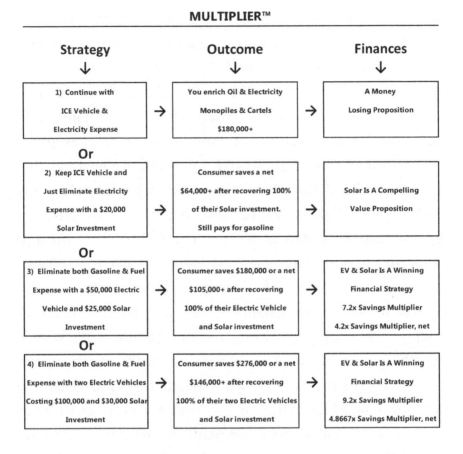

Strategy ↓	Outcome ↓	Finances ↓
1) Continue with ICE Vehicle & Electricity Expense	You enrich Oil & Electricity Monopiles & Cartels $180,000+	A Money Losing Proposition
2) Keep ICE Vehicle and Just Eliminate Electricity Expense with a $20,000 Solar Investment	Consumer saves a net $64,000+ after recovering 100% of their Solar investment. Still pays for gasoline	Solar Is A Compelling Value Proposition
3) Eliminate both Gasoline & Fuel Expense with a $50,000 Electric Vehicle and $25,000 Solar Investment	Consumer saves $180,000 or a net $105,000+ after recovering 100% of their Electric Vehicle and Solar investment	EV & Solar Is A Winning Financial Strategy 7.2x Savings Multiplier 4.2x Savings Multiplier, net
4) Eliminate both Gasoline & Fuel Expense with two Electric Vehicles Costing $100,000 and $30,000 Solar Investment	Consumer saves $276,000 or a net $146,000+ after recovering 100% of their two Electric Vehicles and Solar investment	EV & Solar Is A Winning Financial Strategy 9.2x Savings Multiplier 4.8667x Savings Multiplier, net

Electric Vehicles paired with Solar eliminates energy dependencies. The qualitative information contained in this book, along with the quantitative data results from the Multiplier model, completely validates Elon's vision of being energy positive and empowering the individual as their own utility. This is why Electric Vehicles that are paired with Solar to power both transportation and electricity consumption needs becomes an Electric Vehicle and Solar Savings Multiplier making them a winning financial strategy!

Try recovering your ICE vehicle costs and getting it for

free? You could. You would have to buy real estate above an oil reserve, drill for the oil, process and refine your own fuel, build an underground storage tank to store the gas along with building your own fuel dispenser. Don't forget you also would need to build your own electrical power plant to avoid electricity expenses so as not to also enrich Electricity companies. It seems like a heck of a lot of trouble to go through. Nevertheless, there are those diehard ICE burners who just love coal rolling (look it up on YouTube) and love hating on Elon and Tesla.

Going with the aforementioned ICE option strategy will undoubtedly take months or years, not to mention a whole lot of real estate and construction for the oil rig tower, refinery, gas pump, and electricity rig setups. It may look just a bit silly with them in your front or backyard, not to mention the difficulty in getting your State and City to permit them. On the flip side, you will be the absolute envy of every ICE coal roller.

Solar is simple to install, and the Sun's energy is always free! Never waste free! Always capitalize on free! Elon is correct! The Sun's free energy allows a consumer to become energy positive and empowers them as their own utility. When done correctly, solar eliminates both gasoline and electricity dependencies. All you need to do is buy an Electric Vehicle and Solar Energy System and have real estate to install it as a roof mount, ground mount, or a carport. My family takes great comfort knowing we are no longer enriching the Energy Monopolies. We are in complete control of our own energy and not dependent on others.

We became energy positive, empowered as our own utility.

How do you like the Multiplier so far as a winning financial strategy? Are you still not convinced, or are you confused?

Okay. You are still not convinced. I will provide another analogy in case you still have reservations of not going with an Electric Vehicle and Solar strategy. I look at the economics of Solar similar to being a homeowner versus a lifelong renter. Solar is an investment as much as a house is an investment. Paying rent is not an investment and is a lifelong expense much like paying for gasoline and electricity expense. As a homeowner, you will own your home free and clear after paying off your loan. Once the mortgage is paid off, it immediately frees up cash flow for investment or savings purposes. The same can be said for Solar. Solar is a powerful investment that can be used as an effective hedge against both rising gasoline and electricity costs when paired with an Electric Vehicle. Once your Solar investment is paid off, you will own it free and clear just like a home. Once paid off, it also immediately frees up cash flow for investment or savings purposes. Being hedged with just Solar on a stand-alone basis to eliminate electricity is good. However, being double hedged with both an Electric Vehicle and Solar Energy System to eliminate both gasoline and electricity is great!

Let's examine renting. Consumers who are renter's will be subjected to lifelong rental increases charged by their landlord. Rent expense will always increase by some unknown amount and will always be a drain on a consumer's cash flow. The same can be said for gasoline and electricity cost increases that are controlled by the Monopolies. If a consumer chooses to remain dependent on their energy sources, they will be forever subjugated to whatever price increases they command. Much like a renter subject to lifelong rent increases, being dependent on Monopolies for energy will always be a drain on one's cash flow. Ever-rising rent costs are always a financial burden. Add to this a lifetime of ever-rising gasoline and electricity expenses and it adds up to a sizeable triple hit to a consumer's finances.

To recap, when done correctly, Electric Vehicles and Solar are a winning financial strategy. The choice is yours. Choose a life of being a double hedged energy independent badass or a double energy dependent dumbass? You know what's cool about being a double hedged energy independent badass? Telling people who drive ICE vehicles your Tesla's and Solar are free compliments of Elon's Secret Master Plans. But wait, it gets even better. Electric Vehicles and Solar are not only a winning financial strategy, they are also a winning environmental strategy as a result of significantly reducing or eliminating one's long-term CO_2 emissions. Now, tell people not only are your Tesla's and Solar are free, but that you are in fact **Triple Hedged™**, protected against future gasoline and electricity expenses and emitting CO_2 pollution by using Elon's strategy of being energy positive. With an Electric Vehicle and Solar energy positive strategy, you become a **Triple Protected, Triple Hedged, Triple Badass™**. Elon Musk was right in his Secret Master plans. Thank you, Elon!!!

To recap, imbued within Elon's Secret Master Plans reveal an underlying strategy enabling a consumer to become triple hedged by being energy positive as illustrated below:

TRIPLE HEDGED™

When you engage in an energy positive strategy with Electric Vehicle and Solar you are **Triple Hedged™** against:

1) No gasoline expense.
2) No electricity expense.
3) No CO_2 pollution.

You become a **Triple Protected, Triple Hedged, Triple Badass.**™

With an Electric Vehicle and Solar strategy that is NOT energy positive you are still triple protected to a certain degree against:

1) Rising and changing gasoline prices.
2) Rising and changing electricity prices.
3) Emitting CO_2 pollution.

If you are still not convinced or are confused with all the metrics, you may wish to consider the online Multiplier model available at www.theenergymatrix.com. It addresses all these dynamics and calculations in a clear layout for a more informed financial decision. Ask yourself if it's worth spending a few minutes to model out your transportation and energy strategy to save tens, if not hundreds of thousands of dollars in the future? Walk through it with someone who is a financial professional that you are comfortable with. The model addresses multiple elements and is (hopefully) organized to provide financial perspective in a straightforward manner.

It should be noted that the Multiplier is contingent upon many variable factors:

1) Being Energy Positive – The key to maximizing the Multiplier assumes being energy positive. This means producing more energy than your household or business and Electric Vehicle's need.

2) Solar Energy System Cost – Spending $75,000 versus $30,000 on a Solar Energy System will directly impact the Multiplier outcome. The more expensive the Solar Energy System, the lower the Multiplier result.

3) Solar Energy System Space and Orientation – The more space available the better position a consumer is in to add the number of solar panels needed to become energy positive. Additionally, solar panels are best optimized in a south facing position in the U.S. If a consumer only has space available in a north, east or west facing position more solar panels would be needed as solar irradiance position is not optimized.

4) Solar Energy System Type – Solar Energy panels are not equal. Some have higher efficiency ratios than others and lower degradation rates than others. A cheaper, lower costing Solar Energy System with a 10-year warranty will allow you to recover your costs sooner than a more expensive premium Solar Energy System with a 25-year warranty. However, cheap, and lower costing is not where you want to go. A 10-year solar warranty may only produce up to 20 years of energy whereas a premium 25-year warranty system can produce up to 40 years of energy, resulting in

higher long-term energy savings. Additionally, a lower-end Solar Panel System could degrade faster, resulting in lower energy production yield over the life of the solar panels.

5) The Current State where a consumer primarily resides – Gasoline and electricity is more expensive in California than in Louisiana. The higher the energy prices in one State will allow a consumer to recapture their costs quicker than in a State with lower energy prices. The State's geographic location is also an energy element in the amount of solar irradiance it receives. Temperate and cooler Washington requires additional Solar Energy Panels for energy generation than sunny and warmer California.

6) Electric Vehicle Cost – Spending $150,000 versus $50,000 on one electric vehicle will directly impact the multiplier outcome. The more expensive the Electric Vehicle, the lower the net Multiplier result.

7) ICE Vehicles average miles per gallon (mpg) and annual miles driven – A consumer who commutes 10,000 miles annually and drives a Toyota Prius that averages 55 mpg consumes less gas than a consumer who commutes 20,000 miles annually and drives a Chevy Camaro that averages 25 mpg. A consumer who drives a vehicle that consumes more gasoline incurs a higher expense cost than one who drives a higher mpg ICE vehicle. A consumer who drives more annually and has a lower mpg ICE vehicle would recover their costs quicker than someone who drives less annually and has a higher mpg vehicle.

The online Multiplier model captures many multiple data dynamics in rendering meaningful data outputs. It allows users the ability to change their current energy consumption needs both for electricity and miles traveled in helping to formulate Solar System sizing energy needs along with the Multiplier outcomes.

The following information will aide a consumer in going through the online Multiplier model:

1) Once accessed, consumers select from a drop-down list of the State they reside in. Once selected, the current average cost of their States electricity and regular gasoline fuel costs will auto-fill.

2) **ELECTRICITY tab** – Consumers then need to input their monthly kilowatt-hour (kWh) electricity usage (obtained from their electrical utility bill) over a 12-month time period. This information is needed to determine both the annual kWh electricity consumption and to compute an estimate of the annual electricity costs. This is only an estimate as the State's overall average is used. Electricity costs will vary across multiple Utilities in any given State. Additionally, some Electricity Utilities utilize time-of-use tiered rates. In such instances, the overall electricity rate and cost could be much higher. Lastly, this does not include for possible non-bypassable charges or local municipal taxes.

3) **GASOLINE VEHICLE tab** - Up to 12 gasoline ICE vehicles currently owned or under consideration can be modeled. Select the gasoline vehicle from a drop-

down list, input the vehicle's average miles per gallon (mpg), and the vehicle's annual miles driven. This is to provide perspective on how much monthly and annual gas expense amounts to.

4) **ELECTRIC VEHICLE tab** – Select the Electric Vehicle(s) under consideration as a replacement to the ICE vehicle(s). Once selected, the model will calculate the estimated annual kWh of energy needed for each Electric Vehicle based upon the annual miles input for the ICE vehicle in the GASOLINE VEHICLE tab.

5) **PURCHASE PREMIUM tab** – Input the estimated cost of each respective ICE vehicle in the GASOLINE VEHICLE tab and the estimated costs of each respective Electric Vehicle from the ELECTRIC VEHICLE tab. This is to demonstrate the net premium paid in consideration of purchasing an Electric Vehicle over an ICE vehicle.

6) **SOLAR SYSTEM ANALYSIS tab** – This provides a high-level overview of your annual household or business electricity usage in addition to the estimated annual kWh of energy needed for your Electrical Vehicle(s) transportation. Being Energy Positive means producing more energy than one consumes. This involves oversizing your Solar System size to ensure adequate energy coverage. Bottom line - more energy is always better than less. The model will provide an estimated Solar System size in kW along with the number of panels needed (based on a 360W per panel output). A consumer can choose the number of Tesla Powerwall's to cover energy needs between sunset and sunrise in addition to providing

backup energy protection. An estimated Solar System cost is provided based upon average solar costs for the State selected.

7) **HOME EQUITY LOAN tab** – Demonstrates payment details for financing 100% of the Solar Energy System with a Home Equity Loan. A consumer will input the loan interest rate along with the number of years to pay back the loan. The information displays a change to a consumer's overall cashflow based upon their current monthly gasoline and electricity expenses net of their Solar loan used to eliminate these expenses. The monthly Solar loan payment will replace the monthly gasoline and electricity expense payments.

8) **EXPENSE ANALYSIS tab** – This provides a 10-year lookback and a 40-year forward-looking projection of both the combination of gasoline and electricity expenses. **10-Year Lookback** – This information provides an estimated lookback of a consumer's cumulative electricity and/or gasoline expenses and provides an estimate into how much a consumers cash went to enrich the Oil and Electrical Utility Monopolies over the last 10 years. This information is derived from the current annual electrical consumption in addition to current annual miles driven. This information is then applied to historical cost averages of these energy sources to compute the estimated historical expense. Cumulative means increasing by successive additions. For example, in 2021 if you paid $4,500 in combined electricity and gasoline expenses and you paid $4,000 in 2020, your cumulative amount over a two-year period would be $4,500 + $4,000 = $8,500. A 10-year lookback for these cumulative

expenses would be the sum of 2021+2020+2019+2018+2017+2016+2015+2014+2013 +2012 for all the years combined. **40-Year Forward-Looking Projection** – This is an estimated projection of electricity and/or gasoline prices on a cumulative basis. This information provides perspective into how much of a consumer's cash in the future will go to enrich the Oil and Electrical Utility Monopolies over the next 40 years should a consumer not choose a strategy to become Energy Positive by adopting an Electric Vehicle and a Solar Energy System. Recall, most solar panels are warrantied for 25 years, however premium solar panels can produce energy for up to 40 years. A 40-year forward-looking cumulative analysis starting in 2023 would be the sum of all years from 2023 thru 2062. Remember, Electricity and Gasoline energy prices will generally always increase over time.

9) **MULTIPLIER RESULTS tab** – Highlights the future estimated cost savings (Return on Investment) over a 20, 25, 30, 35 and 40 year time period of a consumer who chooses a strategy of being Energy Positive with an Electric Vehicle paired with a Solar Energy System versus those consumers who continue with the strategy of gasoline vehicles and electricity bills. The Expense Analysis and Multiplier Results tab reflect the absolute true and tragic horror of those who continue to engage in being bound to the Energy Dependent Monopoly Model. Consumers can enrich the Monopolies or transition to the Energy Independent Savings Multiplier Model.

The online Multiplier data reveals the future financial costs of continuing with paying for gas and electricity. You can choose to save this money in the future or choose to enrich the Monopolies. Choose to be a lifelong energy independent financial victor or choose to be a lifelong energy dependent financial victim? The choice is yours. This represents the other 50% of the true horror that awaits consumers' finances. Queue vomiting on the scale of the Exorcist! Talk about the ultimate gasoline and electricity expense black pill barf-o-rama! Yes, a horrific financial barf multiplier awaits those who continue to drive ICE vehicles and pay electricity utilities! Do you hear that screaming? It's your future finances at risk. The Multiplier represents the strategy of avoiding these expenses by becoming energy positive through energy independence. It captures Elon's vision for consumers to become energy positive, empowered as one's own utility through the combination of Electric Vehicles and Solar.

Start popping those green jelly beans! In other words, you should start your online orders with Tesla for its S3XY Solar lineup sooner rather than later. Consumers around the world will soon awaken to The Energy Matrix and all the financial and environmental benefits of Trinity. Once they understand why Elon wants consumers to be energy positive, empowered as their own utility, they will see the possibility of being energy independent. Once they utilize the Multiplier model, they will most likely go full-blown energy independent. Millions of consumers globally will start popping those vibrant green jelly beans too! What flavor are these jelly beans you ask? It's a trio of succulent watermelon, crisp green apple, and juicy pear to tie into the trio of Electric Vehicles, Solar, and Battery Energy Storage solutions. This trio is named TRINITY TREATS (of course), a symbolic sweet treat for your finances, Humanity and for Earth. I'm no doctor, but I can safely recommend the moderate consumption of green jelly beans to realize the sweet financial and environmental benefits of Tesla's S3XY Solar product solutions. Trinity is financially and environmentally sweet indeed! (Note: If you are a diabetic or prone to dental cavities, please skip the jelly beans. Proceed directly with Tesla's Solar product solutions as the benefits are still sweet financially and environmentally.)

Consumers, seeing their future finances at risk through the optics of the Energy Independent Savings Multiplier Model, will opt for green jelly beans and awaken on a mass level, thereby delivering a triple fatal blow to gasoline cars, the Oil Industry, and Electrical Utilities. This just doesn't accelerate the world's transition to sustainable energy; it will make the world go ludicrous speed towards the Energy Trinity S3XY Solar ecosystem solutions created by Elon Musk and Tesla.

But wait, Tesla owners say, "Why not plaid versus ludicrous?" Ask any Tesla owner, and they will agree that the only thing more ludicrous than ludicrous is going plaid! Ludicrous speed will eventually give way to plaid speed under the Trinity ecosystem. I will reveal how and why the world will go plaid in the chapter – TESLA HAS NO COMPETITION.

What's your Electric Vehicle and Solar Savings Multiplier?

I highly encourage you to evaluate what your financial position is and calculate what your savings multiplier would be with a qualified professional. It may not be possible to recover 100% of your Electric Vehicle and Solar Energy System cost. We will explore this, and I will show you how our family is getting ours all cost recovered, thereby making them free.

12

YOUR ELECTRIC VEHICLE(S) AND SOLAR CAN BE FREE

"Some people don't like change, but you need to embrace change if the alternative is disaster." – Elon Musk

Your Electric Vehicle(s) and Solar can be free. Why not "is free"?   There are multivariable attributes in determining whether your Electric Vehicle(s) and Solar "can be free."

The Multiplier models this information by allowing a user to stress the model in applying their own dynamic "What-If" scenarios.   "What-if" scenarios are always used in quantitative financial analysis to provide insight into multidimensional data dynamics.   For instance, what if a consumer drives a Toyota Prius that gets 55 mpg, drives only 3,000 miles per year, the gasoline in their state costs $2.75 a gallon, and their average electricity bill is only $50 a month? It would take much longer to recover one's costs from an Electric Vehicle and Solar Energy Systems acquisitions strategy standpoint compared to someone who drives a Ford Raptor that gets 10 mpg, drives 25,000 per year, the gasoline in their state costs $4.75 a gallon and has an average electricity bill of $300 per month.

Further, in such an example, if the owner of the Prius decided to replace their car with a $135,000 Plaid Tesla Model S and a $80,000, 5kW Solar Energy System who also lives in Seattle, Washington, they might not fully recover both costs compared to the Ford Raptor consumer who replaces their car with a $50,000 Tesla Model 3 and a $30,000, 9kW Solar Energy System who lives in Los Angeles, California. Who would pay $80,000 for a 5kW system? Yikes! You'd be surprised at how shady and convincing some Solar Salespeople are! I've seen it happen all too often. Do yourself a huge favor and check initial Solar pricing on Tesla's website as a starting point. Tesla is transparent with their pricing, just as they are with their Vehicles. What you see is what you pay. There is no salesperson gaming the figures making a commission at your financial expense. If Tesla is listing a 12kW system with one Powerwall Battery for $40,000 and you get a competing Solar quote for a similar system priced out at $80,000, then you know there is a pricing disconnect. You should be very suspect on the competing Solar quote. Also, be careful of the old "bait and switch" where they might quote you something over the internet or over the phone and before you know it the numbers have been reworked to something different from what you were originally advised. I've also seen this happen all too often. Bottom line - buyer beware when it comes to Solar Energy Systems. Yes, Solar will be confusing because it's new and unknown to most consumers. It might be tempting to place your faith in a nice friendly salesperson, however, do your own research. Talk to friends, family, or neighbors regarding their solar experiences. Knowledge is crucial and Solar is where it will all go wrong if not done properly as will being scammed on pricing, product, and warranties over an incorrect Solar Energy System. Stay away from all YouTube Solar ads. Elon even tweeted on June 7, 2022, stating that YouTube is full of "nonstop scam ads." I agree. Please, just don't go there!

Ok, back to the example. The consumer residing in Seattle receives less solar irradiance/solar insolation than the consumer in Los Angeles. Solar irradiance/solar insolation is a measure of solar energy striking a specific area over a set amount of time. In the Solar industry, this is expressed and measured as kilowatt-hours per square meter per day or $kWh/m^2/day$. The amount of sunshine received is higher and more consistent in warmer and sunny Los Angeles, which is greater than 5.75 $kWh/m^2/Day$ than Seattle at less than 4 $kWh/m^2/Day$, which has a more temperate and cloudy weather system. The same holds true across the coast such as Miami, Florida which boasts more sunny days and warmer weather compared to cooler and temperate Boston, Maine. As a result, Solar Energy Systems will yield different levels of energy production generation depending on their geographic location. The online Multiplier model captures many of these multivariable costs and energy dynamics to provide perspective into how such costs would look over a 20, 25, 30, 35, and 40-year cumulative time horizon.

Take another deep, deep breath. Ready? Breathe in.... and go.... The main key factors a consumer should take into consideration is whether they believe an investment today in an Electric Vehicle and Solar Energy System is worth their financial consideration in providing future price protection by eliminating both gasoline and electricity cost risks for the next 20, 25, 30, 35 or 40 years relative to how much they expect to commute and how much electricity they expect to consume in the future net of their Electric Vehicle and Solar costs relative to if they are able to achieve being energy positive, empowered as their own utility in relation to how much Solar Energy they are able to produce in relation to where they reside geographically ....and breatheeeeeeeeee. Haha! Yes, these are just some of the multidimensional variables which are needed in order to decode Elon's vision, or maybe I'm just daft.

Now going back to the first chapter, I will demonstrate to you our family's strategy of how we are getting $190,000 in Tesla's and Solar for free. Note: None of the purchase agreements reflect state taxes and registration fees. I wanted to reflect the detail in whole pre-tax and fees numbers for a cleaner look. I will start with the actual total purchase agreements of our three Tesla's:

T ≡ 5 L ⊼

Motor Vehicle Purchase Agreement
Vehicle Configuration

Customer	Description	Total in USD
	Model 3	$35,000.00
	Rear Wheel Drive	-
	Black Interior	-
	Red Multi-Coat	$1,000.00
	19" Sport Wheels	$1,500.00
	Long Range Battery	$9,000.00
	Premium Upgrades	$5,000.00
VIN 5YJ3E1E. 8056	Subtotal	$51,500.00
	Destination Fee	$925.00
Reservation	Documentation Fee	$75.00
Order Payment $3,500.00	Order Modification Fee	$0.00
	Vehicle Total	$52,500.00
Accepted by Customer on 4/8/18 9:40 PM		

T ≡ 5 L ⊼

Motor Vehicle Purchase Agreement
Vehicle Configuration

Customer	Description	Total in USD
	Model 3	-
	Model 3 Long Range RWD	$49,000.00
	Rear-Wheel Drive	-
	Premium Black	-
	Pearl White Multi-Coat	$1,500.00
	19" Sport Wheels	$1,500.00
	Premium Interior	-
VIN 5YJ3E1E, 0010	Subtotal	$52,000.00
Reservation	Destination Fee	$925.00
Order Payment $3,500.00	Documentation Fee	$75.00
	Transportation Fee	$0.00
Accepted by Customer on 6/27/2018 10:19:16 PM	Order Modification Fee	$0.00
	Total	$53,000.00

Motor Vehicle Purchase Agreement
Vehicle Configuration

Customer Information

Description	Total in USD
Model 3	$35,000.00
Long Range Rear-Wheel Drive	$8,000.00
Rear-Wheel Drive	-
All Black Premium Interior	
Solid Black	-
19" Sport Wheels	$1,500.00
Premium Interior	-
Subtotal	**$44,500.00**
Destination Fee	$1,125.00
Documentation Fee	$75.00
Transportation Fee (if applicable)	$0.00
Order Modification Fee (if applicable)	$0.00
Total	**$45,700.00**

VIN	5YJ3E1E	7793
Reservation		
Order Payment	2,500.00	
Accepted by Customer on	3/18/2019	
Odometer	50	

99

We were able to realize various savings through tax incentives and cash rebates dropping our $189,000 gross cost outlay to a net of $143,000 as follows:

Tesla Model 3 - Red	$	52,500
Tesla Model 3 - White		53,000
Tesla Model 3 - Black		45,700
Solar Energy System		38,036
Gross electric vehicle & solar expense	$	189,236
Less:		
Federal & State tax and cash incentives		
Tesla Model 3 - Red		(10,000)
Tesla Model 3 - White		(10,000)
Tesla Model 3 - Black		(6,250)
Solar - 30% Federal Incentive Tax Credit		(11,411)
Local utility incentives		
Tesla Model 3 - Red		(450)
Tesla Model 3 - White		(450)
Tesla Model 3 - Black		(450)
Less:		
Employer incentive on Electric Vehicle		(4,000)
Employer rebate on Solar		(1,000)
Solar company rebates		(2,000)
Net Electric Vehicle and Solar Expense	$	143,225

Most consumers fail to realize the annual and cumulative expenses that both fuel and electricity costs have on their finances. The following table represents the actual fuel and electricity costs my family paid over a three-year period in 2015, 2016, and 2017. Approximately $27,000 of our income went to enrich the Monopolies over this time period. This was us being trapped in the Energy Dependent Monopoly Model:

| Gasoline fuel costs | | Actual | | | Total 3 year Expense |
		2015	2016	2017	
Acura MDX - 17 mpg	$	2,644	$ 2,540 $	2,767	$ 7,951
Acura RL - 19 mpg		2,112	1,756	2,042	5,910
Honda Accord - 29 mpg		1,958	1,764	2,059	5,781
Total gasoline		**6,714**	**6,060**	**6,868**	**19,642**
Electricity Costs					
Jan		150	180	193	
Feb		148	178	183	
Mar		133	159	169	
Apr		93	112	130	
May		146	174	191	
Jun		158	171	186	
Jul		239	293	301	
Aug		285	330	344	
Sep		293	341	373	
Oct		147	176	201	
Nov		227	247	268	
Dec		181	217	257	
Total electricity		**2,201**	**2,578**	**2,797**	**7,576**
Total gasoline and electricity expense	$	**8,915**	$ **8,638** $	**9,665**	$ **27,218**

We strategically dumped our three ICE vehicles for three Tesla's and a Solar Energy System. In doing so, we avoided almost $39,000 in actual gasoline and electricity expenses from 2018 to 2021 which would have gone to enrich the Monopolies:

Gasoline and Electricity Expenses avoided by dumping our ICE Vehicles and buying 3 Tesla's and a Solar Energy System

Gasoline fuel costs	2018	2019	2020	2021	Total 4 year Expense
Acura MDX - 17 mpg[1]	$ 2,195	$ 3,350	$ 2,654	$ 3,523	$ 11,722
Acura RL - 19 mpg[2]	1,215	2,253	1,756	2,533	7,758
Honda Accord - 29 mpg[3]	2,164	1,076	1,467	2,409	7,115
Total gasoline	**5,574**	**6,679**	**5,877**	**8,465**	**26,595**
Electricity Costs					
Jan	212	209	211	207	
Feb	193	188	189	168	
Mar	165	164	161	167	
Apr	136	131	132	136	
May	191	204	218	225	
Jun	197	204	217	231	
Jul	328	345	359	361	
Aug	368	369	405	420	
Sep	386	400	430	454	
Oct	208	215	229	236	
Nov	266	260	277	279	
Dec	250	242	258	248	
Total electricity	**2,901**	**2,931**	**3,085**	**3,132**	**12,049**
Total gasoline and electricity expense	**$ 8,475**	**$ 9,611**	**$ 8,962**	**$ 11,597**	**$ 38,644**

1) Disposed of April 2018
2) Disposed of June 2018
3) Disposed of June 2019

The $38,644 in gasoline and electricity expenses we avoided went towards recovering our Tesla and Solar Investments. That $38,644 in savings paid for our 1st Tesla, thereby making it FREE! This is now us being free in the Energy Independent Multiplier Model!

Cash leaving our pockets which would be going into the pockets of the Oil and Electrical Utility Monopolies, are no more. Starting from 2018, our cash is now strategically re-directed towards recovering our Tesla's and Solar Energy System investments as illustrated:

Tesla Model 3 - Red	$	52,500
Federal & State tax and cash incentives		(10,000)
Local utility incentives		(450)
Employer incentive on Electric Vehicle		(4,000)
Net Tesla Model 3 - Red Cost	$	38,050

Fully cost recovered in 4.0 years by 2021

Tesla Model 3 - White	$	53,000
Federal & State tax and cash incentives		(10,000)
Local utility incentives		(450)
Net Tesla Model 3 - White Cost	$	42,550

Fully cost recovered in 8.4 years by 2025

Tesla Model 3 - Black	$	45,700
Federal & State tax and cash incentives		(6,250)
Local utility incentives		(450)
Net Tesla Model 3 - White Cost	$	39,000

Fully cost recovered in 11.9 years by 2029

Solar	$	38,038
Employer incentive on Solar		(1,000)
Solar Company rebates		(2,000)
Solar - 30% Federal Incentive Tax Credit		(11,411)
	$	23,627

Fully cost recovered in 13.5 years by 2031

We factor the recovery of our Tesla's on the front end as they are expected to have an estimated life of at least 15 years (Remember, the average age of vehicles in the U.S. at the end of 2021 was 12 years old). Our Solar Energy System is factored on the backend as it is expected to have an estimated life of 40 years.

By becoming energy positive, empowered as our own utility, we recovered 100% of our 1st Tesla Model 3 in just under 4 years. We will recover 100% of our three Tesla's and Solar Energy System costs in just over 13 years by the year 2031, making everything FREE!!! As previously noted, the detail does not reflect state taxes and registration fees. This was done solely to provide a cleaner look. Yes, these are additional added costs which cannot be avoided and would have brought the total to above $200,000. We can assume just over another year to recover these tax and registration costs too!

Awakening to the Energy Matrix and acting on my strategy has been blissful financial liberation for my family! The expenses we were paying day after day, month after month, year after year were significant. We accepted them because the only other alternatives were walking, biking, or commuter transportation which weren't feasible for our transportation needs. That all changed the moment we became energy independent. By going with Electric Vehicles and Solar, we eliminated that queasy feeling of enriching the Monopolies that were draining our finances day after day, month after month, year after year. It's a simple but impactful financial lesson I taught my kids at a very young age that should apply to everyone: Would you rather choose to enrich others or choose to enrich yourself?

It's just not possible to recover one's costs with an ICE vehicle through another energy means. This further demonstrates why Electric Vehicles paired with Solar are a winning financial strategy.

13

YES, YOUR ICE VEHICLE IS OBSOLETE

"When Henry Ford made cheap, reliable cars people said, 'Nah, what's wrong with a horse?' That was a huge bet he made, and it worked." – Elon Musk

Still not convinced an Electric Vehicle is for you? The actual reality is that all ICE vehicles of today are the horse and carriages of the early 1900s. Consumers may not realize this or want to believe it, but it's true. I didn't want to believe it either.

I grew up as a teen riding Suzuki motocross bikes. I was absolutely hooked on the smells of the two-stroke, 40 to 1 gas to oil mixture as I was to the raging sounds ripping from the exhaust pipes. The cushy suspension setup was like pillows and marshmallows as I tore up boulders and berms. I grew up very much enjoying the smell of gasoline. Yes, gasoline was a wonderful and intoxicating smell that brought me much happiness in my youth. But I did what most people do. I grew up. My teenage life experiences and my needs changed. I graduated from motocross to sport bikes opting for a Suzuki GSXR 750 & 1000. I was riding on

Laguna Seca Raceway at the very mature age of 18. I was a top speed and corner apex junkie. The only way I could get my adrenaline fix was always pushing them to the limits. How quickly could I get to top speed? How fast could I attack an apex and how hard could I roll on the throttle exiting out of it – all without losing it? Miraculously, I survived many harrowing accidents walking away with not only my life and limbs but amazingly, without a single broken bone. I was indeed a reckless lucky cat with nine lives. Those nine lives were being quickly exhausted the more and more I indulged my adrenaline needs. Luckily, cat number 6 called out to me and gave me a complete knock out upper cut. Getting taken out on a top speed run and smacking my head on the pavement was my wakeup call. My Shoei helmet took the full impact of my fall followed by my competition weighted racing leathers. I most certainly would not have survived without them. It was a very rude awakening but surviving this last accident was just the one I needed to snap me out of my top speed and apex junkie habits.

Before I knew it, I got married and started a family. My transportation needs changed as I continued to grow up. I did the whole Honda, Acura, Toyota, and Lexus auto gamut for my family. As the years progressed, I no longer enjoyed getting gasoline nor the smell of it. I grew tired of filling up all of the vehicles at Costco, getting annual smog checks, performing routine oil changes, and haggling at the dealerships when the time came to get a new car. I grew frustrated in dealing with auto dealerships whose service advisors were commission based making service recommendations that weren't warranted. I grew nauseous of the monthly recurring fuel expenses eating away at our household's monthly finances. All of these costs added up cumulatively into thousands of dollars per year. I did what most rational consumers do. Minimize the fuel expense while trying to maximize the mileage. Traveling to Costco to save $.20 per gallon and feathering the throttle to save 2 mpg became just downright irritating.

Then, on March 31, 2016, the whole notion of my auto transportation mindset took a complete 180. March 31, 2016 was the day Elon Musk debuted the Tesla Model 3. Like so many others, I watched the event via the live stream webcast. When the Model 3 was unveiled, I was stunned at the absolute beauty of the overall design of its exterior. It was a cartoon moment – my eyes popped out of my head and my tongue rolled out onto the floor. The sportiness coupled with the fluidity of the design, clean interior with minimalistic aesthetics, and center mounted screen were futuristic, and exciting. The pricing specs were secondary, and just like that, I whipped out my credit card. I happily placed two separate $1,000 deposits securing two reservations. I waited two years before getting an email notification in March 2018 advising me to prepare to take delivery of my first Tesla Model 3. The problem was that I had never driven a Tesla before. The Model 3 I was to receive was one of the earliest productions off the factory production lines. Additionally, none were available to test drive at a Tesla Center. Fortunately, TURO, a peer-to-peer car rental platform, already had a Tesla Model 3 available for rent. It was being rented out by a SpaceX engineer whose strategy was to recoup his costs when he was not driving it. Lucky me! I was hesitant to commit to a full-on purchase without having an opportunity to at least test drive it. I rented the Model 3 for the entire weekend. I figured this would be enough time that I needed to make an informed decision on whether to go through with the purchase or not. It turns out I didn't need the weekend. Nope. I didn't even need 5 minutes. I was sold and hooked the moment I stepped on the accelerator. My family's transportation considerations would forever change from this point forward. Goodbye gasoline cars forever! Hello Tesla! From the delivery to ownership experience, I knew the Tesla Model 3 was going to be an absolute smashing hit! Driving and owning the Tesla freed my mind forever of what transportation should be. I could never drive another ICE vehicle again. They feel so inefficient and darn right slow. Plus, I always felt guilty stepping on the gas when I needed to. All I could think about was the added pollution I was causing being released into the atmosphere, not to mention the added cost when it came to filling up the tank. The engineering and performance quickly had me realize that the Tesla Model 3 would be responsible for ushering in a new automotive era not seen since the automobile was first introduced in the early 1900s. It is that revolutionary. We are witnessing the effects of Tesla's Electric Vehicles on the global auto market today.

Before the invention of the modern-day ICE vehicle, global

consumers of the 1900s primary source of transportation were the horse and carriage. When the modern-day ICE vehicle was introduced, it was met with much laughter, skepticism, and resistance. It's easy to understand why. It's typical human nature to be skeptical of new ideas and resist change. A century ago, people were too afraid to part with their dear old horses, much like individuals today who are too afraid to part with their ICE vehicles. Individuals are comfortable with what is familiar to them and what they have grown up with understanding their whole life.

We are witnessing the next evolution in automotive transportation compliments of Elon Musk and Tesla. By the end of this century, very few ICE vehicles will remain operational on Earth. The few ICE vehicles that will remain will be on display by a handful of museums, much like the few that display horse-drawn carriages of a century ago. Some will remain as artwork collection pieces, some will remain in garages for sentimental reasons, and some will still remain operating at higher equatorial latitudes due to the lack of solar irradiance and cold weather inhibiting a consumer to become energy positive. Ask yourself, how many horses and carriages of a hundred years ago do you see on the roads today as a primary source of transportation? Not many. The same will apply to ICE vehicles in the future.

The modern-day automobile replaced the horse-drawn carriage of a century ago. The mobile smartphones of today replaced the rotary dial phones of 50 years ago. Revolutionary disruptive technologies will always do this. All

ICE gasoline vehicles have been outengineered by Tesla and are simply obsolete. Electric Vehicles emit no noise, are safer structurally, cleaner to drive, have way fewer moving parts, require no oil changes, require less maintenance, and now, with the information and knowledge contained in this book, can be cost recovered through the utilization of solar. Advances in Electric Vehicle transportation and technology evolved through Elon's vision and Tesla's engineering group. The execution of their product lines directly paved the way for gasoline vehicles extinction. As a result, every major legacy ICE auto manufacturer has announced plans to eliminate their fossil-fueled vehicles at a future date and transition to Electric Vehicles.

Trinity will drive mass consumer adoption towards energy independence once consumers read this book and have an opportunity to run different financial strategy outcomes with the Multiplier model. They will come to realize that Elon's vision in his Secret Master Plans were correct. Electric Vehicles and Solar upend the Monopolies energy control. Consumers now have a choice to become energy independent, eliminating two expensive financial dependencies – fuel and electricity while also significantly reducing their CO_2 emissions.

Yes, believe it. Your ICE vehicle is obsolete. This is not just an isolated event happening in the U.S. It is happening globally. All 195 world countries have agreed to the 2015 Paris Climate Accord, a legally binding international treaty to reduce greenhouse gas emissions to mitigate climate change. All the legacy ICE auto manufacturers are now planning for it. Legacy ICE auto manufacturers have all made announcements to phase out ICE vehicles in pursuit of alternative energy transportation means, mainly Electric Vehicles, given Tesla's safety, successful global sales figures, performance track records, engineering, technology, and proven reliability. Yes, believe it. Your ICE vehicle is obsolete much like the horse and carriage and the rotary phones of decades past.

Now let's explore how to get you set up in Tesla's S3XY Solar product solutions.

14

YES, YOU CAN AFFORD AN ELECTRIC VEHICLE AND SOLAR ENERGY SYSTEM

"If you think back to the beginning of cell phones, laptops or really any new technology, it's always expensive." – Elon Musk

You say you don't have the financial means to pay for a Solar Energy System upfront, let alone an Electric Vehicle right now?   That's a valid concern as a majority of consumers' resources are stretched.

"Who can afford an Electric Vehicle?" It's what many view in their misunderstanding of Electric Vehicles. This goes back to the misunderstood upfront costs versus the more significant long-term financial risk exposure that both gasoline and electrical costs represent to consumer finances. I believe most consumers can afford an Electric Vehicle and a Solar Energy System. We sold our ICE vehicles and used the proceeds as down payments towards our Tesla's and financed the remaining balance. We paid for our Solar Energy System in full.

Most consumers are coming to realize that gasoline vehicles aren't cheap these days either. Gasoline vehicle prices are at all-time record highs and pushing the upper limits of pricing in 2022, coming in at $47,000 according to a January 14, 2022, Kelly Blue Book article, *Average New Car Price Tops $47,000*.

A financial education update is needed to understand that while Electric Vehicles may be more expensive upfront, an investment made today to go with an Electric Vehicle paired with a Solar Energy System that powers both transportation and electricity consumption needs provides significant financial cost savings over the long run. This strategy will provide long-term protection when extrapolating for how much the elimination of both gasoline and electricity expenses will add up to in the future.

Sadly, we live in an era wherein many consumers can only think about their current monthly expenses rather than preparing for their long-term financial future. It is totally understandable given the recent inflation hits across the entire consumer spectrum. Nonetheless, consumers need only to reflect on their childhoods and recall Aesop's fabled story – *The Ant and the Grasshopper*[6]. It has been passed down over two millennia from generation to generation. Everyone knows the story:

> One bright day in late autumn a family of Ants were bustling about in the warm sunshine, drying out the grain they had stored up during the summer, when a starving Grasshopper, his fiddle under his

---

[6] https://read.gov/aesop/052.html

arm, came up and humbly begged for a bite to eat.

"What!" cried the Ants in surprise, "haven't you stored anything away for the winter? What in the world were you doing all last summer?"

"I didn't have time to store up any food," whined the Grasshopper; "I was so busy making music that before I knew it the summer was gone."

The Ants shrugged their shoulders in disgust.

"Making music, were you?" they cried. "Very well; now dance!" And they turned their backs on the Grasshopper and went on with their work.

The overall moral of the story being, work hard, and invest smart in the present to ensure both a safe and secure long-term future. This same principle should be applied to consumers' finances when evaluating which energy strategy to pursue over the long term.

A consumer should pose the following long-term question to themselves and to their families: Choose a life being forever dependent on the Monopolies energy or choose a life of being forever energy independent, empowered as their own utility?

Be the Ant - Work hard and make the right choices today. Make good long-term investment decisions in the present to realize and reap the fruits of one's hard-earned efforts, which will be bountiful in the future. They are beholden to none and accountable for their own actions. Energy Independence would be a strategy adopted by the Ant: Securing their future by eliminating long-term financial risk from ever rising energy costs.

Conversely, be the Grasshopper – Don't work hard or make the right choices today that are needed in the future. Don't make sound long-term investment decisions in the present, leading to future financial uncertainty causing stress, and in a worst-case scenario, financial dependency on others. Being energy dependent would be a strategy adopted by the Grasshopper: Remaining dependent on the Monopoly's for gasoline and electricity thereby enriching the Monopolies and placing their family's future long-term finances at risk of higher and higher energy costs.

Grasshoppers make excuses. Ant's work towards solutions. Which one will you choose to be?

The Ant, of course! Excellent choice!

Now let's examine how to get you in an Electric Vehicle and Solar Energy System.

Paying cash and in full for an Electric Car and a Solar Energy System is a nice option for those few who can afford to do so. However, you may not want to pay full, in cash, even if you have the means to do so because of other investment opportunities. You may want to save that extra cash and possibly put some of it to use once you read my TESLAMAXSFAST chapter. These are some possible strategies to consider (in no particular order):

1) Sell or trade-in your ICE vehicle quickly and use those proceeds towards an Electric Vehicle and Solar Energy System. As more and more consumers awaken within the Energy Matrix, they will come to understand the long-term benefits of being Energy Independent. This will result in used cars flooding the markets. ICE vehicles will be worth significantly less in the future than they are today. The more you can get for your trade-in, the more this will aid in offsetting your new Electric Vehicle and a Solar Energy Investment purchase.

2) Consider a Home Equity Loan for the Solar Energy System. Many homeowners are currently flush with equity. It may be prudent to tap into that equity rather than taking out other types of loans. The loan may be tax-deductible for the solar panels, which can make it more ideal than other loan options. Consumers may find that funding a Solar Energy System with a home equity loan with a 7 to 10-year repayment term will not change their overall monthly cash flow. Their monthly home equity loan payment will instead replace their former gasoline expenses and electricity bill. This financial element is a component in the Multiplier model. But what happens if a consumer moves? Many consumers do. I address this in the next section. Every consumer's financial position is unique. Therefore, they should consider reviewing with a certified financial professional and tax advisor before pursuing any large financial transaction and tax strategies.

3) Ask Parents for assistance. Parents may have the financial resources to help. Alternatively, it can be approached as an advance from an inheritance. Why wait until their passing as they can see how their children or grandchildren will benefit today by eliminating their energy dependency and financial risk exposure for the next 40 years?

4) Consider taking out a larger, longer-term auto loan. Many consumers finance their ICE vehicles. These same consumers who opt for an Electric Vehicle would still need an auto loan. Consumers should evaluate selling their ICE vehicle(s) and use the proceeds towards a new Electric Vehicle or Electric Vehicles. No consumer wants to take a larger or longer-term loan. However, the elimination of gasoline expenses will offset some, if not most of the increased loan balance. Eliminating $300 or $600 a month in gas would be re-directed as an offset to a higher auto loan and towards a Solar Energy System. This financial element is also captured in the Multiplier model. Electric Vehicles have fewer operating maintenance costs than traditional ICE vehicles. We have incurred zero maintenance costs over a 3- and 4-year ownership period of our Tesla's. We do not have to pay for smog checks or oil changes – FOREVER! These ICE costs add up incrementally over time.

5) Consult with a professional financial advisor. This book, the Multiplier model, data, and concepts may be just too overwhelming for some consumers to understand. A professional financial advisor may have

a better way to present the data and formulate a financial strategy to fit a consumer's budget.

Okay, but your current home isn't your "forever home." It's understandable. Many consumers frequently move before settling in on their final home. What happens if a consumer installs a Solar Energy System and decides later to sell their house or business and move? Will they lose their solar investment? Not necessarily. Solar panels are portable, whereas solar roofs are not. This is why I focus discussions only on solar panels due to their portability. You can simply hire a qualified Solar installer to disconnect service and remove the panels to include in your U-Haul or with a professional moving transportation company for relocation. You can then arrange to have another licensed solar installer install the panels on your relocated home. Alternatively, you can factor it in as a purchase premium item on top of the home sales price if you wish to forgo the trouble. It is a transaction that would have to be fully weighed on the merits of the costs versus the benefits of relocation.

As I previously mentioned, Solar is where it could all go wrong for consumers in the future. It will be a very costly mistake. So much so, that I will need to write a complete book about just Solar and the need to get it done right. You would think Solar Professionals would know however, I had four Solar Professional representatives build out my system and they were all wrong based on my research and calculations. I crunched the data and product metrics to ensure I was well informed beforehand and was surprised by just how incorrect their recommendations were.

There are many infomercials on T.V. and YouTube about how consumers can get paid to get solar or how a consumer can get solar for zero down. They are everywhere. Just don't go there. If you are tempted by their questionable and too good to be true advertising, be very wary and skeptical of their claims. These are primarily consumer data collection sites to which their personal information will be gathered and sold to pushy solar salespeople who will bombard consumers with multiple phone calls and emails. You will know if they start asking for your email and contact information before allowing you to view or model solar information on their websites. Proceed with a high degree of caution should you go this route as it could be very, very costly in the future should your Solar system spec's come out wrong.

There are Solar representatives at Costco's. Sadly, the 4 or 5 different representatives I spoke with at various locations knew absolutely nothing about their own products. They didn't know if they manufactured their own panels, or where they were made. They didn't know what their efficiencies or degradations were. The list goes on and on. They knew absolutely nothing and demonstrates how wrong solar can go.

Consumers should avoid Solar Leases and Purchase Power Agreements or PPA's. There is a lot of information widely available on Solar PPA's on the internet. Therefore, I won't rehash something I wouldn't recommend. Generally, there are more long-term downside risks that outweigh any upfront benefit a salesperson would be cross-selling for this financial arrangement.

I researched Solar extensively before moving forward with my plans. Sadly, the four different Solar sales individuals I met with were all wrong about their Solar recommendations. I knew what I wanted in advance of meeting with them. A Solar salesperson's main objective is obviously to sell their product. My experience was they approached it by being sensitive to a consumer's cost needs rather than a system that was truly right for my needs. Much like auto salespeople they can find any price to match your financial needs. However, this is a totally misguided approach. Pursuing such a strategy can leave one open to financial risk exposure in the future.

One Solar sales representative we met received a phone call and his phone caught my eye. What caught my attention were the 78,000+ unread emails. Yes! 78 thousand unread emails! I can only imagine what those customers must have been going through in trying to reach him. Needless to say, his Solar offer was the lowest of the four. Although he was a really nice guy, those 78,000 unread emails said it all. No way! No how! No matter how good the price was. Going cheap will get you what you pay for. So yes, Solar is where it can and will all go wrong if not done properly. It's a subject befitting a whole separate book.

15

AN ENERGY REVOLUTION IS COMING

"Man has the power to act as his own destroyer - and that is the way he has acted through most of his history."
– Elon Musk

A new revolution is coming. It will be an energy revolution of power over the people versus the idea of free power by the people as their own utility.

The future of oil and electricity energy commodities will, without question, be one of the most significant news topics of this decade as the movement towards energy independence gains momentum. Social and media energy wars will be waged with the proliferation of Electric Vehicles, Solar Energy, consumer awareness of The Energy Matrix, Trinity, the Electric Vehicle and Solar Savings Multiplier model, and the notion of being Triple Hedged.

Oil and Electrical Utility Monopolies will do anything and everything in their financial power to preserve their income streams and profits. Just how much of their future revenue is at stake? Over $ 100 trillion in just the U.S. alone over the next 40+ years. What about globally? Globally, $300+ trillion

USD equivalent is at stake over the next 40+ years. Yes, TRILLIONS!

$300,000,000,000,000+

The Monopolies stand to lose $300+ trillion while global consumers stand to save $300+ trillion. Make no mistake. The Oil and Electricity industry's main objectives will be to keep global consumers hooked and remain on their Energy Dependent business models as consumers seek to exponentially transition towards energy independence. Monopolies have spent well over a century of investment and control forged with extensive global relationships with governments, state agencies, and political favors. The idea of a consumer not being financially dependent on their energy sources will not be allowed by any and every means necessary.

The financial clout, political influence, and reach of their wealth and power are not to be taken lightly whatsoever. Politicians, State, and Local Government Leaders, World Leaders, Technology Companies, Influential Media Personnel, Celebrities, YouTube Influencers, energy company personnel, Educators, certain well-funded advocacy groups, and news organizations will all be parroting energy misinformation under the guise of the Monopolies and Cartels dictation. They will use each of these delivery channels to exert their influence over global consumers. The Energy Matrix, Trinity, the Multiplier model, being Triple Hedged, and the notion of any consumer energy independence will be challenged at every level, just as Elon and Tesla have. Multi-pronged, well-coordinated misinformation media and advertising campaigns will be launched on such a massive scale that it will make the last decade of mainstream media misinformation seem like an episode of Sesame Street.

There is a black silver lining to this, but sadly, it comes at a long-term financial and environmental cost. As word spreads of The Energy Matrix and Trinity, energy prices will eventually be managed down to drown out the news about the current high global gasoline and electricity costs. The Monopolies approach is that most consumers are like goldfish with a 3 second memory. They tend to forget about the high prices if energy costs are managed down. It won't hurt as much on the consumers' pocketbooks. This will be done to appease the masses, with an underlying strategy to keep them hooked on their lucrative business model. Their goal is for Consumers to consume black pills, forever remaining in their Energy Dependent Monopoly nightmare. This is precisely what drug dealers do to addicts. They keep drug prices low enough to make addicts repeat buyers. The addicts remain in a lucid state as an always returning customer with prices low enough to get their cheap fix. Imagine if they increased their prices? The addict might be inclined to look elsewhere or stop using their product and go cold turkey. Drug dealers will have none of it. The addict's money is their money. The same mentality applies to the Monopolies.

Still don't believe that energy Monopolies and Cartel's control supply, pricing and that the future of energy costs will go higher?

Here are some recent quotes by Monopolies and Cartel's on the subject:

As prices of oil broke $100 a barrel in early 2022, the Saudi energy minister, Prince Abdulaziz bin Salman, rejected calls to increase oil production noting that renegotiating quotas between OPEC members risks bringing volatility to oil markets rather than stabilizing them.

"The kingdom is not on the same page with the U.S. currently. We all know they (OPEC) are not ready to cooperate with the U.S. to calm the market."

In December 2021, Saudi Arabia's energy minister also warned traders against shorting oil after the collapse of oil prices after the Thanksgiving holiday, saying OPEC+ could react quickly to any fall in prices. "Thanksgiving was a Thanksgiving Day for the speculators," the minister said. "But let them dare to do another Thanksgiving. They will be ouching like hell."

Electrical Utilities are also flexing their monopoly price increases with Edison International Chief Executive Pedro Pizarro stating in April 2022, "We're obviously very focused on trying to manage electric costs as much as we can, but let's also recognize that there will be more pressure on electric costs." This stemmed from an article noting that Electrical utilities are making big spending increases to update the aging power grids. Translation? Expect significant increases in future electricity rates. Rates to which the consumer has no say over and must pay in order to keep the lights on.

Speaking of aging and less dependable power grids, the problem will likely get worse before it gets better. Frequent, sustained power outages have occurred with increasing frequency and size in the U.S. over these last two decades. Federal data by the U.S. Energy Information industry shows fewer than two dozen major disruptions in 2000. In 2020, the number surpassed 180. Curt Morgan, CEO of Vistra Corp., which operates the nation's largest fleet of competitive power plants selling wholesale electricity, said he is worried about reliability risks in New York, New England, and other markets. "Everything is tied to having electricity, and yet we're not focusing on the reliability of the grid. That's absurd, and that's frightening."

Lastly, the February 2021 Texas Power Crisis resulted in over 200+ deaths caused by a series of severe winter storms that had knocked out a significant portion of Texas's electrical grid. This electrical grid outage affected some 4.5 million homes and businesses for several days. Texas's electrical grid is separate from U.S. national grids in order to avoid federal regulation. This separation did not allow for the state to import electricity from other states during the outage. Texas is one of the few states that allows for consumers to shop for wholesale electricity rates directly. Electrical wholesalers immediately increased their production rates which resulted in consumers being charged for thousands of dollars' worth of electricity.

One customer named Scott Willoughby was both lucky and unlucky as a result of this storm. He was lucky because his lights stayed on during the storm. Sadly, he was unlucky because he received an electricity bill to the tune of $16,752! "My savings is gone," said Mr. Willoughby, a 63-year-old Army veteran who lives on Social Security payments in a Dallas suburb. He said he had nearly emptied his savings account so that he would be able to pay the $16,752 electric bill charged to his credit card — 70 times what he usually pays for all his utilities combined. "There's nothing I can do about it, but it's broken me." He was not alone. Many thousands of other consumers were hit for thousands of dollars on their monthly electricity bill. For customers whose electricity prices were not fixed and were instead tied to volatile wholesale prices, the energy rate price spikes were egregiously insane!

Electric Vehicles and Solar Energy removes the variability of risks when it comes to your finances: future energy price volatility and future energy price increases. Do you want to look back 40 years from now and then conclude you just enriched the Monopoly's tens or hundreds of thousands of dollars? Do you want to have life regrets wishing to go back in time thinking if you could just do it all over again? What kind of regrets? Regrets thinking back that you could have used those cost savings to pay off your mortgage quicker, pay for your kid's education, or taken the family on epic vacation trips. I don't think any person wants those kinds of regrets in the future. My family and I certainly do not and will not have those regrets. Do take this into consideration when listening to the upcoming energy dependence versus energy independence misinformation campaigns.

It will be very easy to filter out all the forthcoming energy revolution misinformation. Just remember these three simple fundamental questions: Choose to enrich the Monopolies or choose to enrich yourself? Choose to be forever dependent on their energy subject to their price increases for life or choose to be forever energy independent hedged against price increases for life? Choose to be a lifelong energy financial victim or choose to be a lifelong energy financial victor?

Now for the real boring stuff. If you are wondering what kind of daft person could come up with decoding two aspects of Elon's Secret Master Plans by building Multiplier models, unlock the Multipliers, build their own website, write and self-publish this book, and come up with the book cover art then read on at your own risk. It that all sounds like a bit much to digest you can proceed straight to the TESLAMAXSFAST chapter and skip these next few chapters.

16

CONNECTING THE DOTS AND PACIFIC COAST BANKING SCHOOL

People should pursue what they're passionate about. That will make them happier than pretty much anything else.
— Elon Musk

Ok, so you didn't skip and go straight to the TESLAMAXSFAST chapter. I debated with much reservation whether to include the next two chapters in this book. I ultimately decided to include it solely to address a basic question I suspect Quants, Artists, and a few others might want to know. How did I decode two key elements within Elon's Secret Master Plans? Additionally, how was I able to unlock and build quantitative financial Multiplier models from both secret Master Plans that deliver a triple fatal blow to gasoline ICE cars, the Oil industry, and Electrical Utilities for the entire United States in just two days, build and design my own website, write and self-publish this book that describes everything qualitatively and quantitatively, and come up with the book cover art? These are reasonable and fair questions.

If you are curious, you can continue on. If not, go ahead and skip it. Full disclosure: it's a complete tangent from this book's sub-premise on what was just covered, "Why Electric Vehicles and Solar are a Winning Financial Strategy." It focuses instead on decoding Elon Musk's Secret Master Plans and the book cover art. Yes, it is the title of the Book. Why would I have reservations about writing about it? Well, it was all very simple but with a lot of complexity which I know doesn't make any sense. The complexity lies within all the elements of my professional background that lend merit to it all. Full disclosure: It comes off as self-serving, a bit narcissistic, highly technical, a bean counter, a quant's, and an art lover's dream however, every non-financial and non-art person's worst nightmare. This is not my intention whatsoever. It's massively confusing to anyone without a finance or accounting background or who appreciates art. It's extremely dry, long-winded and was done solely to understand just a little bit about me and how I came to decode two key elements of Elon's Secret Master Plans, build, and unlock the Multiplier models, build, and design the Energy Trinity Website, write and self-publish this book, and design the book cover art. What type of person would even attempt to do such things? Read on if you are curious. However, in all fairness, I am once again advising that you may wish to skip it. It will be a confusion multiplier for the non-finance savvy and non-art lovers. Gulp!

I have never been employed by Tesla or had a career in the Automotive or Solar Industry. My lifelong career has been in the Financial Services Industry. My street credentials are embedded in both quantitative finance and technical accounting. I cut my teeth and rolled up my sleeves working in multi-million and multi-billion dollar organizations as a senior executive financial officer over the last 20+ years, analyzing and presenting tremendous amounts of internal and external data. This internal and external financial data was for C-Suite teams, board of directors, auditors, and regulatory examiners.

In 2017, I was accepted into Pacific Coast Banking School (PCBS), a premier national graduate school of Banking located in Seattle, Washington. Pacific Coast Banking School, in partnership with the Foster Graduate School of Business at the University of Washington, is a three-year, master's-level management extension program for senior and executive officers in the Banking Industry.

Not anyone can simply enroll in this program. Only the most elite and promising next generation of America's top banking financial leaders are nominated and evaluated for acceptance into this program. Once nominated, each applicant is then thoroughly screened to ensure they possess the necessary education, financial skillsets, and banking knowledge depth to handle the demanding rigors of the program.

At the beginning of my first year in the program, it was noted and observed that I was a key student with a majority of the answers to the Professors questions across many broad banking and financial subject matters. Many school peers wished to partner with me and have me lead assignment-based projects and discussions. I demonstrated the ability to convey technical banking, finance, and accounting subject matter in a concise, straightforward, and succinct manner. This became even more apparent as the now retired and former CEO of PCBS, Dave Enger, approached me after completing the first year's session. Dave let me know that due to my coaching and insight provided to fellow peers during class discussions, that two fellow professors with a Ph.D. and an MBA were not invited back to the school. As a result, I developed a very favorable following amongst my banking peers based on my level of knowledge and leadership skills. I was a key subject matter expert to my peers in the area of Finance, Accounting, and Regulatory knowledge.

Over the course of the three-year program, I was required to attend a wide range of Banking and Finance courses ranging from Dynamic Leadership, Managing Bank Financial Performance, Enterprise Risk Management, Business Ethics, and Global Finance. I was also required to complete multiple in-depth written extension assignments encompassing Dynamic Leadership, Credit Risk Management, Managing Bank Financial Performance, Practical Finance for Bankers, Asset Liability Interest Risk Management, and Enterprise Risk - each requiring 15 – 30 pages of subject matter. Additionally, an original management report on the subject of Mergers and Acquisitions totaling 25-50 pages was required in my final third year.

It was during my second year at PCBS in 2018 that I accidently decoded two key elements within Elon's Secret Master Plans from building the Multiplier model. It stemmed from my professional experience and a second-year paper I needed to write on Asset Liability Management, Interest Rate Risk, and Enterprise Risk Management. I'm including it for both reference and educational purposes as I built the model leveraging off my experience in modeling and managing Asset Liability and Interest Rate Risk exposure. It was a key component in formulating the concept of the Multiplier model as it came about simultaneously while writing this school report. I felt it prudent to include so that readers understand some fundamentals of the model and my thinking approach towards implementing financial solutions to maximize income while mitigating future financial risk when it comes to dealing with multidimensional data dynamics. I developed the Multiplier for consumers as the goal of any individual, family, and business is to maximize income and liquidity while mitigating or eliminating both current and future financial risk through strategic planning and execution.

Here was my report submission at Pacific Coast Banking School on Asset Liability Management, Interest Rate Risk, and Enterprise Risk Management assignment:

Fair warning: about 99.8% of you will most likely not understand this. Most Bank CEOs and some CFOs even have a hard time understanding it. It encompasses multidimensional data dynamics and was a key element in formulating the Multiplier models. For confidentiality purposes, I removed all references to the financial organization's name, including certain data metrics. Queue the confusion multiplier...

Income Simulation

XXXX Bank's balance sheet at June 30, 20XX was asset sensitive. Its interest earning assets will reprice faster than is interest-bearing liabilities. As a result, the Bank will experience an increase in net interest income (NII) in a rising rate environment. The magnitude and degree to which the Bank is asset sensitive is derived from and illustrated by its Static – Repricing GAP Balance Sheet Report. The Bank has a majority of repricing assets in excess of maturing liabilities in a majority of various stratified time intervals and has a total positive +$XXX.X billion cumulative rate sensitive assets to rate sensitive liabilities position.

The Bank's net income will increase from base static scenario if rates rise by the following: +11% or $XXX +100 basis points (bps), +23% or $XXX +200 bps, +34% or $XXX +300bps and +45% or $XXX +400bps. The Bank's net income will decline from base or rates unchanged if rates fall by the following: -14% or ($XXX) -100bps, -27% or ($XXX) -200bps, -36% or ($XXX) -300bps and -41% ($XXX) -400bps.

The Bank has policy limits established for both Net Interest Income and Economic Value (EV). Net Interest Income policy limits are identical for rates both up and down as follows: -8% +/-100bps, -15% +/-200bps, -20% +/-300bps, -20% +/-400bps. At June 30, 20XX, the Bank was not in compliance with policy for Net Interest Income in the -100bps, -200bps, -300bps and -400 bps scenarios. The out of policy exceptions for Net Interest Income are a result of the Bank's ability to keep its deposit costs down and unchanged relative to the rising rate environment. The Bank's weighted average cost of deposits excluding (DDA) was XXX bps at June 30, 20XX and modeled rates cannot be shocked below zero. No action plan is warranted at this time as to do so would require increasing deposit rates, which would adversely impact the Bank's net interest margin and profitability.

The pros of extended income simulations are that they provide greater Net Interest Income prospective longer than a one-year time horizon. The cons of this are that the balance sheet is dynamic and will change throughout the year which will affect the future Net Interest Income outcome.

The pros of ramp rate income simulations are that they are more in alignment with historical FOMC policy. The cons are it does not take into consideration instantaneous rate shock exposure.

The pros of instantaneous shock scenarios are they provide quantitative measure to demonstrate changes in Net Interest Income across various interest rate scenarios. This provides management with perspective on how the balance sheet can be strategically positioned relative to the direction of interest rates. The cons are it does not take into consideration longer-term interest rate risk exposure.

Two main core deposit assumptions impact income simulation results: deposit decay and deposit betas. Deposit decay takes into consideration deposit life on nonmaturity deposits withdrawal behavior. The deposit decay impacts the Net Interest Income simulation as it leverages off historical company specific data to quantify deposit migration. The deposit beta refers to how much an institutions interest-bearing products rates will change in relation to market rate changes. The deposit beta impacts the Net Interest Income simulation results by demonstrating changes to interest on deposit expense in relation to various interest rate stress-shocked scenarios. The higher the institutions deposit beta the more deposit expense will increase leading to compression in Net Interest Income.

Economic Value

The Bank's Economic Value profile at June 30, 20XX is liability sensitive. Its Economic Value decreases faster in a declining rate environment. The magnitude and degree to which the Bank's Economic Value is liability sensitive is derived from and illustrated by its Static – Repricing GAP Balance Sheet Report.

The Bank's base static Economic Value scenario will increase if rates rise by the following: +3% or $XXX +100 basis points (bps), +7% or $XXX +200 bps, +10% or $XXX +300bps and +12% or $XXX +400bps. The Bank's static Economic Value will decline from base or rates unchanged if rates fall by the following: -6% or ($XXX) -100bps, -12% or ($XXX) -200bps, -19% or ($XXX) -300bps and -20% ($XXX) - 400bps.

Economic Value sensitivity policy limits are identical for

rates up and down as follows: -8% +/-100bps, -15% +/-200bps, -20% +/-300bps, -20% +/-400bps. At June 30, 20XX, the Bank was in compliance with policy risk limits for EV sensitivity in all interest rate scenarios.

Deposit beta's and deposit decay assumptions also impact Economic Value by varying the cash flows that make up the very basis of Economic Value. Changes in deposit betas and deposit decay will impact both the earnings and cash flows to which are modeled and captured for Interest Rate Risk and Economic Value purposes.

Economic Value is derived by determining the present value of all asset cash flows less the present value of all liability cash flows. In a rising rate environment, the price of assets will generally go down greater than the increase in the value of its liabilities for an asset sensitive positioned balance sheet.

The pros of Economic Value are that it provides a quantitative measurement on interest rate risk as it takes into consideration the change of value of assets and liabilities under different rate scenarios. Additionally, it takes into consideration all cash flow and maturity components on the Bank's interest sensitive assets and liabilities. As a result, it provides insight into the Bank's longer-term Interest Rate Risk position. Economic Value is a longer-term interest rate risk measurement that allows management to assess how strategic decisions made today effect the Bank's financial position and future earnings. It does not measure income over time. Unfortunately, the cons of Economic Value are that Economic Value shocks the balance sheet instantaneously in calculating overall value at risk, however, provides no detail on the timing or future rate risk. Future cash flows can be difficult to quantify for deposits and other financial instruments that have no maturity because these types of products have uncertain duration and cash flows. Additionally, the discount rate assumptions used may not be accurate in deriving the present value of interest earning assets and interest-bearing liabilities.

Total Return

Our institution does not use a total return report. The Bank's income simulation shape is directionally in harmony with its Economic Value. The interest rate scenarios that present the most risk to the organization is the rates down scenarios -100 bps, -200 bps, -300 bps, -400 bps. The magnitude and degree to which the Bank's Net Interest Income is asset sensitive is derived from and illustrated by its Static – Repricing GAP Balance Sheet Report. Net Interest Income and Economic Value will both decline in a rates down scenario. The shapes for Net Interest Income versus Economic Value are not at odds. The institution is asset sensitive for Net Interest Income and liability sensitive for Economic Value. Total return provides a clearer picture of risk versus reward than income simulation and economic value as it combines both aspects of factoring in all cash flows while also accounting for the passage of time. Total return also provides a clearer picture to risk versus income simulation and Economic Value as it quantitatively provides dollar volume specific impacts to net income over multiple rate scenarios. It does this by measuring all the risk versus rewards of assets and liabilities in different combinations across various interest rate stress scenarios by demonstrating the difference between current versus future income.

Application

XXX Bank uses interest rate risk data to assist with asset and liability duration in combination with interest rates offered on both loans and deposits in positioning the Bank's balance sheet to maximize net interest income and shareholder value while mitigating financial risks.

XXX Bank views Asset Liability Management as a valued decision-making function that adds significant value. It allows our institution the ability to model static and dynamic scenarios through means of various inputs and multiple "What-if" strategies. The Bank makes decisions to mitigate risk and enhance shareholder value utilizing Asset Liability Interest Rate Risk Management in running multiple "What-if" scenarios for both loan and deposit products by rate and term. For example, what happens to Net Interest Income in a rates up or down scenario if we consider purchasing a $XX billion multifamily loan pool with a 6.55% start rate, a floor of 4.5%, no ceiling, tied to Prime with an average spread of 1.55% funded by $XX billion in new money market funds at 0.75% and $XX billion 1-year CDs at a rate of 1.25%. Such a projected model with multidimensional data dynamics provides insight as to whether those strategies in repositioning the composition of interest earning assets and interest-bearing liabilities are in alignment with policy risk limits and the desired risks and reward outcomes for management and its shareholders. The Bank uses this information strategically before changing the composition of the balance sheet to determine what the impact is to Net Interest Income. Additionally, that the Bank remains within policy risk limits and is adequately rewarded for the risks assumed.

A current strategy to improve the overall Asset Liability Management position of our institution would be to deploy excess cash (liquidity ratio of 27% 2QXX) to purchase a $XX billion AAA rated CRE loan pool yielding 5.00% net of premium. The asset mix repositioning will enable the Bank to bolster its interest earning asset yield while still maintaining a sufficient 15% liquidity ratio.

A. Executive Summary

I used XXXXX's Component Risk Score Model as the basis for the risk assessment methodology for XXX Bank's CAMELS (Capital, Asset, Management, Earnings, Liquidity, Sensitivity) component risk assessment. The methodology and scoring are to align my understanding and to allow for the instructor/grader of this assignment ease of cross reference when analyzing my risk score assessments in relation to XXXXX's risk scoring model. XXX Bank's overall risk score was 200 after taking into consideration all the composite key risk indicators (KRI's) scoring.

C	A	M	E	L	S	
Capital Risk	Asset Quality Risk	Management Risk	Earnings Risk	Liquidity Risk	Sensitivity Risk	Overall Risk Score
38	23	40	38	33	28	200

XXX Bank is based in XXX and was established in 20XX. Its FDIC peer group is 5 – Insured Commercial Banks having assets between $300 million and $1 billion, comprised of 1,220 institutions. The aforementioned data, KRI's and the following executive summary is based upon the following combination of data: 1) Current composite scoring; 2) XXX's 20XX Annual Report, FDIC Call Reports, and trended FDIC Uniform Bank Performance Report data; 3) XXX's data metric's relative to its FDIC peer group, bank industry norms and FDIC guidelines.

Recommendations to improve XXX Bank's risk profile:

• Preserve capital and equity by slowing asset growth by focusing in on loan quality rather than loan quantity and curtail dividend distributions. XXX's 11.16% risk-based capital is approaching the 10% "well-capitalized" threshold.

• While its asset quality ratios are very good now, the economy has been on a decade of economic expansion. Lending history has demonstrated that during an economic down cycle, an institution's best loans were made during the worst economic times and its worst loans were made during the best economic times. XXX should focus on shoring up underwriting guidelines, borrower debt service coverage ratios with a concentration on loan quality over quantity. The trended loan growth demonstrates a large concentration of unseasoned loans that have yet to experience an economic down cycle which can place future asset quality at risk.

• Focus on growing core deposits as liquidity has been managed down to 15%. We are currently in a rising rate environment. This will translate into a change in the deposit mix and the institution may experience a disintermediation of deposits if it does not focus on growing its core deposit base should it not be able to price deposits both timely and competitively.

• Focus on non-interest income by re-evaluating other fee income products. XXX's has trended down 10% year over year from 20.14% in 20XX down to 10.54% in 20XX. This was due to a $333K other than temporary impairment on one corporate bond coupled with a $150K decline in mortgage broker fee income.

• Focus on reducing all general and administrative costs by re-evaluating benefits cost structure and price negotiate with all vendors. XXX's efficiency ratio of 71% is above the peer group average of 64%.

B. CAMELS (Capital, Asset, Management, Earnings, Liquidity, Sensitivity) Component Risk Assessment

Capital Risk

Key Risk Indicator	12/31/2017 Ratio	Risk Score	Score Rationale
Risk Based Capital	11.16%	44	Above FDIC PCA 10% threshold for Well-Capitalized and above capital with phased-in conservation buffer of 9.25% however, below peer group average of 15.35%
Leverage Ratio	8.48%	32	Above FDIC PCA 5% threshold for Well-Capitalized however, below peer group average of 10.32%
Tier 1 Risk-based	10.29%	37	Above FDIC PCA 8% threshold for Well-Capitalized and above capital with phased-in conservation buffer of 7.25% however, below peer group average of 14.26%
Average score		38	- LOW

Capital Risk represents the amount of equity available relative to how risk weighted that equity is against an institution's assets (Risk Based Capital – Assets are stratified and categorized per FDIC Call Report guidelines). A low capital risk ratio can be indicative of either high risk (growing assets relative to equity) or a reduction in shareholders equity caused by distributions or losses sustained through earnings. The more equity available, the more that institution is able to absorb potential losses and weather cyclical economic downturns. Therefore, an institutions capital position relative to fully risk weighting assets on its balance sheet gauges and assesses the institutions overall level of Capital Risk.

After reviewing XXX Bank's key risk indicators (KRI's) noted above, the current risk score assigned for the Capital Risk component I arrived at was 38 – Low. My assessment takes into consideration XXX's capital levels/ratios in relation based on two metrics: 1) FDIC Prompt Correction Action Levels and 2) additional capital conservation buffers mandated under Dodd-Frank that were effective January 1, 2015 and phased in annually through 2019. The phased in capital ratios with the required conservation buffers for 2017 are as follows: CET 1 = 5.75%, Tier 1 Capital = 7.25%, Total Capital = 9.25%. The Leverage ratio is not subject to a conservation buffer to which 4.00% is the minimum Leverage ratio required. XXX's KRI's are above the requirements in both metrics formulating the basis of my assessment of their overall level of Capital Risk.

Asset Quality Risk

Key Risk Indicator	12/31/2017 Ratio	Risk Score	Score Rationale
Adjusted NPL/Total Loa	0.08%	23	's is below 1% and also well below peer group average of 0.49%. Non-performing loans to total loans represents how well an institution is collecting payments from its borrowers. Non receipt of agreed upon payment terms can lead to a loan becoming non-performing. A high level of non-performing loans is indicative of elevated credit stress.
NPA/Assets	0.06%	20	's is below 1% and also well below peer group average of 0.57%. Nonperforming assets represents NPL + OREO. OREO represents loans an institution acquired through foreclosure with intend to sale. OREO loans are written down to LOCOM based upon most recent appraisal less estimated costs of sale. High levels of NPA's to total assets is indicative of worsening credit stress.
Adj NPA + Adj Lns 90PD/Tang Equity + LLR	0.71%	25	AKA the Texas Ratio - measures the ability of an institutions total equity position to absorb nonperforming assets and other real estate loans. This ratio is very important during economic down cycles. Institutions approaching 100% are at risk of not having sufficient capital to cover credit losses. 's is below 1% which is very low.
Average score		23	- VERY LOW

Asset Quality Risk represents how well an institutions loan portfolio is performing relative to current market conditions and changes. Asset quality is supported by prudent and sound underwriting guidelines within the scope of an institutions credit policy limits, the knowledge and skillset of the loan team and loan committee, and diversification of loans so as to not have industry or geographic concentrations. Asset quality is further enhanced with ensuring adequate borrower debt service coverage ratios supported by primary and secondary sources of repayment coupled with adequate collateral protection.

After reviewing XXX Bank's KRI's noted above, the current risk score assigned for the Asset Quality Risk component I arrived at was 22 – Very Low. My assessment takes into consideration XXX's asset quality levels/ratios in relation to its peer group averages. XXX's KRI's are below peer and industry norms formulating the basis of my assessment of their overall level of Asset Quality Risk.

Management Risk

XXX's Management Risk stance on a trended basis is viewed as moderately aggressive to aggressive. In reviewing the Uniform Bank Performance Report trended data, XXX deployed excess capital to fund asset growth. This is evident in their equity to average assets ratio dropping from 9.70% in 20XX to 8.45% in 20XX. Risk weighted assets grew from $199.7 million in 20XX to $323.6 million in 20XX, a 62% increase or 16% CAGR (compounded annual growth rate).

Although asset quality ratio levels are below peers, the growth in the loan portfolio reflects continued expansion in the current economic recovery cycle and possibly looser underwriting standards. These loans are not vintage and are unseasoned which have the potential to place future asset quality at risk of exposure during any possible economic down cycle.

XXX's net interest margin has shown signs of compression going from 4.11% in 20XX to 3.73% in 20XX. This paints an interesting picture as previously booked loans that matured were replaced at lower offering rates. However, one would also expect to see a corresponding decline in the cost of funds of deposits as time deposits would have also matured and be replaced with lower cost sources of funds. XXX may be growing loans at the expense of yield with lower offering rates resulting in lower Net Interest Income and relatively static net earnings.

Liquidity has been managed down from 33% in 20XX down to 15%. XXX also shows trended increases in reliance of wholesale funding and net non-core funding dependence. Although I rated XXX a low level of liquidity risk as of 12/31/20XX, a point in time. The trended combination of the aforementioned can place XXX at a higher degree of liquidity risk in a rising rate environment. A rising rate environment can lead to liquidity outflows as depositors seek higher yields and put downward pressure on the net interest margin as the deposit mix may shift from MMA into higher costing time deposits.

After reviewing XXX Bank's key risk indicators (KRI's) noted above, the current risk score assigned for the Management Risk component I arrived at was 40 – Low.

Earnings Risk

Key Risk Indicator	12/31/2017 Ratio	Risk Score	Score Rationale
ROAE	6.89%	43	's is below the peer group average of 9.68%. 's lower ratio is indicative of a higher cost structure relative to the generation of its net interest income + other income. While double digit ROAE is ideal in terms of delivery shareholder value, the risks to ROAE must be understood so that short term economic gains do not come at the expense of long term risk or
Efficiency ratio	70.94%	40	's is above the peer group average of 64.32%. The efficiency ratio represents how much in expenses an institution is incurring relative to the net interest income + other noninterest income it generates. 's higher ratio is indicative of a higher cost structure relative to the generation of its net interest income + other income as it spends $0.71 for every $1.00 it earns. A higher efficiency ratio will have a negative impact on earnings.
Net Interest Margin	3.73%	30	's is below the peer group average of 3.85%. A lower NIM ratio can be attributable to a combination of how the balance sheet is structured with duration and embedded options such as floors and ceilings in addition to the pricing of offering rates on both loans and deposits. Compression in NIM results a reduction of earnings unless operating expenses run parallel to that compression.
Average score		38	- MODERATE

Earnings Risk represents an institution's ability to generate profits to support capital growth relative to how well the balance sheet is structured and managed from an interest rate position on both loans and deposits, its underlying loan credit quality complemented with how well the institution manages its expenses.

XXX experienced a sharp decline in ROAE from 10.95% in 20XX to 6.87%. This was attributable to a decline in other noninterest income year over year per XXX's 20XX Annual Report. Other noninterest income went from $1.2MM in 20XX to $699K for 20XX, a decline of $508K or 42% per their Call Report. The decline in other noninterest income is attributable to a $333K OTTI (other than temporary impairment) on a corporate bond they took a charge on coupled with a $150K decline in mortgage broker fee income.

After reviewing XXX Bank's KRI's noted above, the current risk score assigned for the Asset Quality Risk component I arrived at was 38 – Moderate. My assessment takes into consideration XXX's Earnings Risk/ratios in relation to its peer group averages. XXX's KRI's are at odds directionally with its peer group and industry norms in formulating the basis of my assessment of their overall level of Earnings Risk.

Liquidity Risk

Liquidity Risk represents an institution's ability to satisfy client withdrawals with the availability of liquid assets it has to deploy. Managing liquidity risk requires the utilization of cash flow projections based on sources and uses of funds in order to meet current and future client deposit needs without materially impacting the banks operations or financial condition. Liquidity risk is mitigated by institutions access and availability to fed funds facilities with correspondent banks, available borrowing capacity from the FHLB and brokered deposits as primary and secondary sources to fund liquidity.

After reviewing XXX Bank's KRI's noted above, the current risk score assigned for the Liquidity Risk component I arrived at was 33 – Low. My assessment takes into consideration XXX's Liquidity Risk ratios in relation to its peer group averages at point in time 12/31/20XX and not based on historical trended liquidity positions.

Sensitivity Risk

Key Risk Indicator	12/31/2017 Ratio	Risk Score	Score / Rationale
One Year Gap/Assets	5.35%	30	This ratio represents all repricing assets over all repricing liabilites at one year out as a percent of assets. A positive rather than negative ratio indicates Aquesta's Balance Sheet is structured as Asset Sensitive and posititioned to grow NII in a rising rate environment to which are are currently experiencing. The higher the ratio the greater the growth in NII in a rates up scenario.
One Year Cumulative Repricing GAP	$ 21,673	30	This ratio represents all repricing assets over all repricing liabilites over one year expressed in dollars. Positive rather than negative indicates Aquesta's Balance Sheet is structured as Asset Sensitive and posititioned to grow NII in a rising rate environment to which are currently experiencing.
Long term assets/Total Assets	19.63%	23	A lower ratio indicates a lower level of asset duration risk. A shorter duration risk is preferred so as not to be exposed to economic down cycles. Institutions with long asset duration exposure have more sensitivity risk in instances of inverted yield curves.
Average score		28	- LOW

Sensitivity Risk reflects an institutions net interest income impact and exposure of its interest earning assets to interest bearing liabilities in combination with the re-pricing and maturity time horizon of those balance sheet instruments stressed in various interest rate scenarios both rates up and down +/- 100bps, +/- 200bps, +/- 300bps, +/- 400bps. The degree to which and institutions net interest income changes relative to changes in interest rates allows perspective on structuring financial instruments on the balance sheet to maximize earnings and shareholder value while mitigating risk.

After reviewing XXX Bank's KRI's noted above, the current risk score assigned for the Sensitivity Risk component I arrived at was 28 – Low. My assessment takes into consideration XXX's Sensitivity Risk ratios and figures in relation to the current interest rate environment. The FOMC has clearly expressed its desire to increase rates to ensure that 2% inflation remains in check relative to the continued economic expansion of the U.S. coupled with record low unemployment. XXX Bank has a positive one-year GAP profile and is positioned to grow net interest income due to the current rising rate environment.

The aforementioned was my report submission related to Asset Liability Management, Interest Rate Risk, and Enterprise Risk Management. Yes, I know, it was painfully boring, confusing, and dry. Hopefully, future generations of bankers will be able to leverage some of the information from my report to use within their institutions. Well, that was just a sliver of what a Financial Quant's world is like in working with multidimensional data dynamics. We model out multiple scenarios and formulate strategies to yield advantageous results. It's like rocket science, but instead of exploring the universe, it's exploring how to maximize key performance metrics across all business lines.

My professional experience of the aforementioned, coupled with acquiring three Tesla's and a Solar Energy System, provided the collective foundation in formulating and building the Multiplier model. Asset Liability and Interest Rate Risk incorporates many multidimensional data dynamics in assessing how changes to various balance sheet item structures layered with what immediate and future changes to interest rates have on net interest income. Such information provides insight into how to best optimize an organization's financial profile to mitigate immediate and long-term risk while maximizing financial outcomes. On a personal simplistic level, it would be correlated as strategically assessing one's overall financial position and how various strategy changes could impact that financial position both immediately and over the long-term. That information would then be used to one's advantage to maximize the best returns relative to the risks assumed.

I built the Multiplier model for anyone to use in the U.S. to provide perspective into their financial future based on multidimensional energy data dynamics. My goal in the Multiplier model is to demonstrate the elimination of both future gasoline and electricity prices over a 20, 25, 30, 35, and 40 year time horizon and a projection of what those energy prices could be in the future at its most simplistic level. This is necessary as good financial strategies are ones that incorporate mitigating cost risks. Great financial strategies are ones that can eliminate multiple cost risks indefinitely. The combination of Electric Vehicles and Solar represents tremendous complementary benefits and synergies in eliminating three key long-term risks: 1) Fossil-fuel dependencies, 2) Electricity dependencies, 3) Environmental harm caused by CO_2 emissions.

Okay, so I'm a Bean Counter, a Quant, and a Tactical Strategist. I know..... boring, sleeper, and yawn. However, I use this bean counting, quant, and tactical strategic mentality skillset to my advantage. Numbers aren't just numbers to me. I am a financial and strategic tactician with a keen sense of cost and risk mitigation focused on profit maximization. Numbers are my trade tools that enable me to formulate immediate and long-term strategic plans and execute on those plans in order to drive and maximize financial results. Here is evidence of that in action at PCBS:

Pacific Coast Banking School also has two capstone programs in the third year, LEADERSIM (Leader Simulation) and BANKSIM (Bank Simulation). These simulations bring the student's entire PCBS experience into focus, wherein they work and compete with one another in teams as they experience what it is like to manage their own banks. The 2019 third-year session was comprised of 225 of America's brightest Banking and Finance professionals that were segmented into 40 teams consisting of eight Banking Communities. All the teams were presented with an identical $800 million troubled Commercial Bank that was not profitable, was illiquid, and had an overall regulatory CAMELS composite score of 4. The acronym "CAMELS" is a Bank Supervisory Rating System by the Federal Deposit Insurance Corporation and refers to six components of a bank's condition that are assessed during each regulatory review cycle: **C**apital adequacy, **A**sset Quality, **M**anagement, **E**arnings, **L**iquidity, and **S**ensitivity to market risk. Ratings are assigned for each component in addition to the overall rating of a bank's financial condition. Ratings are assigned on a scale from 1 to 5. Banks with ratings of 1 or 2 are considered to present few, if any, supervisory concerns, while banks with ratings of 3, 4, or 5 present moderate to extreme degrees of supervisory concern, with 5 being the worst or most extreme financial concern.

In BankSim, students were advised that the prior management team was responsible for the institution's poor financial performance. As a result, all were dismissed. Our objective was clear: Each group was brought in as the replacement management team to turn around the poor operating and financial performance of the failing bank. Each team was comprised of five to six individuals who were elected to the following roles – CEO, CFO, CCO, Treasurer, Credit Admin, and Retail Deposit Officer. Each team was tasked with turning around the Bank's performance with a goal of earning the highest profitability, highest stock price, and best overall Enterprise Risk (CAMELS) scores.

I was honored to have my team members elect me as the Bank's CEO. Even before BankSim officially started, I had already strategized and formulated an action plan to restore the Bank to profitability and strengthen its regulatory risk scores. The Bank Financials revealed years of escalating financial losses. Additionally, the Bank had a negative liquid assets ratio, had a fed funds purchased to equity ratio greater than 100%, an efficiency ratio of 89%, was marginally capitalized, and was liability sensitive positioned in a rising rate environment leading to further compression in its net interest margin. My strategy to free up liquidity, reposition the balance sheet for rates up, strengthen the efficiency ratio and reduce assets to shore up the Bank's capital base to bolster return on equity all worked in order to both restore profitability and elevate the Bank's overall Enterprise Risk composite scores.

The BankSim course is designed to simulate a two-year financial period compressed into a two-week timeframe. Each day of simulation is designed to represent an entire quarter's financial simulation. Each team was provided with vast amounts of historical and projected quantitative financial data: Balance Sheet, Income Statement, Cash Flow, Liquidity position, Asset Liability position, Interest Rate-Swaps, Loan Credit Quality, Investment Securities, Loan, and Deposit Funding projections, Loan and Deposit interest rates, Capital Ratios, along with projected economic data such as inversion of the 2 and 10-year treasury curve. Each day teams were required to pour over dozens of pages of financial data with a key objective to turn around the bank's financial position – all within a two-hour timeframe. Once the data was scrubbed, each team was required to input various components into the BANKSIM model that they believed would render the best financial and regulatory outcomes. The financial and regulatory results of each team's actions would then be revealed the following day. Each day teams were required to re-scrub all new historical and projected quantitative financial data: Balance Sheet, Income Statement, Cash Flow, Liquidity position, Asset Liability position, Loan Credit Quality, Investment Securities, Loan and Deposit Funding projections, Loan, and Deposit interest rates, Capital Ratios, along with projected economic data.

Teams were required to name their institutions. We named our troubled bank VSG Bank. Let's just say the name VSG Bank is an inside joke. Those who graduated from the 2019 program in our banking community are the only ones in the know.

Each day, every team's financial results were posted for all to see in Paccar Hall. Each day, everyone would rush to see which team's decisions elevated them to the top position for that day. One day a team can be on top and the next day their decisions could place them at the bottom. This was all driven by each team's daily decisions, modeled inputs, and simulation results.

The vast amount and level of quantitative data to decipher in such a short amount of time was highly stressful for students. Individuals were encouraged to be put into roles they did not already occupy as their main profession. For example, my partner Katy was a district branch manager who was selected for the role of CFO. John, who was a Chief Credit Officer, was selected for Deposits and Funding. Zayne was a Financial Data Analyst and was selected for the Deposits role. David was a branch manager who was selected as the Chief Credit Officer, and Vincent, who was also at a branch level, was selected as the team's Treasurer (Note: none of these are their real names for confidentiality purposes). Being in unfamiliar roles was purposedly designed to be stressful for a reason. The stress under pressure in unfamiliar roles was to incorporate the dynamic aspects of LEADERSIM. How each individual would perform in their communications and work cohesively as a team is a dynamic element. Their collective leadership skills honed under the program were designed to ensure each member is learning how to work cohesively as a team under stressful conditions in driving the overall financial performance based on the decisions they would make collectively; All real-world situations that leaders face every day.

As the team's CEO, I never sweated a single detail nor stressed at all the data, nor at the team dynamics, nor at the challenges, and competition that lay before us. This was a role I was prepared for. As a career big data Quant, financial, and strategic tactician leader, I always had massive data dumped on me with the goal of not only financially reporting it, but to develop meaningful data at a succinct level to optimize and drive immediate and long-term financial performance. Preparing U.S. Securities and Exchange Commission Forms S-1's, DEF 14-A's, 10-Q's, 10-K's, and other regulatory filings along with Board Financials, Company Earnings Releases, and Annual Reports over these past 20+ years allowed me to quickly hone in on key data points necessary to succinctly teach and lead my team on how to optimize operating financial performance. My professional experience also encompasses technical finance and accounting from researching and implementing complex accounting pronouncements to top down, bottoms up, and zero-based budgeting/forecasting, to Asset Liability and Interest Rate Risk management, equity award accounting, and hedge accounting. Leveraging Monte Carlo dynamic modeling simulations along with Black Scholes modeling intrinsic value incorporating pricing and rate volatility are also all elements specific to my data modeling background. My background also encompasses collaborating with Big-4 accounting teams, Investment Banks, legal counsel, and government regulators. The culmination of all this experience allowed me to size up the multitude of quantitative data to immediately formulate a strategic action plan to drive and optimize our Bank's performance metrics. I would then communicate my thoughts with the team sharing my insights and rationale in order to execute my vision.

These actions, day after day for the next two weeks revealed the results of our team's strategic efforts. I led our team to finish 1st place overall in our Banking Community in achieving the highest net income, highest stock price, and best Enterprise Risk scores over a two-year economically stressed simulation period. My team was also the only team that did not incur any significant regulatory violations/penalties throughout the two-year simulation period.

We formally presented our strategy and results to the rest of the Bank community teams at the conclusion of the competition. The following were the strategies I instructed my team to act on that were deployed and executed on:

Thank you for attending this year's 20XX annual shareholders meeting. I am very excited to share with you the strategies and efforts undertaken and achieved by our new executive team. I am Neo, President, and CEO of VSG Bank. I would like to introduce you to the seasoned executive team I assembled two years ago at the beginning of 20XX to turn around the financial performance of VSG:

John, our Deposits Officer. Katy, our Chief Financial Officer. David, our Chief Credit Officer. Vincent, our Treasurer and Zayne, our Credit Administrator.

Our strategy at the very beginning encompassed measures to firstly, enhance each aspect of VSG's enterprise risk profile and secondly, restore it to operational profitability. After a careful assessment of the Bank's fourth quarter 20XX poor financial condition, the following issues were immediately addressed to cut continued losses inherited from the prior management team:

1) Return VSG to a positive liquidity position.
2) Shored up capital by reducing VSG's balance sheet.
3) Re-structured and repositioned VSG's balance sheet for the rates up environment.
4) Reduced the Bank's position in Fed Funds Purchased.

We successfully achieved the execution of these four measures during the first quarter of 20XX.

I led and advised my team to execute on the followings strategies we implemented for the first year's simulation run:

Finance and Accounting - We immediately went to the capital markets and successfully raised $12 million in additional capital to ensure a well-capitalized Prompt Correction Action (PCA) designation. Additionally, we took immediate gains on sales of securities and loans totaling $1.2 million to offset shoring up the allowance for loan losses which was underfunded by $4.8 million.

Treasury – We purchased up to $100 million in fixed-rate investment-grade securities to substantially bolster the Bank's liquidity profile.

Deposits – Pricing action coupled with the deposit base needed was challenging given the economic environment. We priced our deposit fees in accordance with our peers and our rates in alignment with national rates. We match price-funded security purchases with FHLB long-term borrowings to secure funding and locked in our asset to liability GAP position.

Credit – The prior management team's credit decisions appeared to have loosened their underwriting standards. We inherited a portfolio of newly originated loans that were unseasoned and had yet to experience an economic downturn. History indicates the best loans are made during the worst economic times, and the worst loans are made during the best economic times. As a result, we believed it to be prudent to adopt a defensive strategy for bolstering VSG's overall loan risk profile.

Credit Admin – We curtailed lending choosing to bring on higher-rated loan credit deals. We adjusted pricing and product mix appetite for portfolios with the greatest amount of return with the least amount of risk.

The aforementioned strategies worked successfully for our Bank. We stopped years of hemorrhaging losses and ended 20XX with a small profit of $100,000. We ended with the highest stock price at $47.92, had the highest dividend payout, and highest risk-based capital ratio. Our overall Enterprise Risk (CAMEL) score was immediately elevated from a 4 to a 2.

Our strategies for 20XX were focused on maximizing operational profitability with targets set for a return on average equity of 18% and return on average assets of 1.15%.

The following strategies were deployed for 20XX:

Finance and Accounting - Economic signals going into the end of 20XX appeared mixed with spreads between the 2 and 10-year Treasuries showing evidence of a sharp contraction and an inversion. We adopted a three-prong strategy to hedge an economic downturn. We went long on fixed-rate loans, initiated interest rate swap positions, and reduced headcount across all business units.

Treasury – We initiated, stacked, and staggered in $50 million increments to receive fix, pay variable interest rate swap positions. We incurred one regulatory violation of $100,000 for not properly obtaining regulatory approval to buy back 6% or, in our misunderstanding, input six shares of stock. We corrected this and did receive approval to execute a buyback of 100,000 shares in the open market. This was done to boost stock share price by improving equity book value.

Deposits – Pricing action coupled with the deposit base needed was still challenging given the economic environment. We positioned our liabilities with more favorable variable rate pricing to ride down the economic cycle in order to preserve our net interest margin. We then locked in long-term fixed-rate deposits and FHLB borrowings at the bottom of the economic cycle to allow for our net interest margin to expand during the economic recovery.

Credit – The prior year's strategy of enhancing the overall loan risk profile paid off. Nonaccruals and charge-offs were manageable. With a strong risk profile, we pivoted and shifted into an expansive mode at the bottom of the economic cycle while also bolstering our Net Interest Income.

Credit Admin – We continued to adjust pricing and product mix appetite for portfolios with the greatest amount of return with the least amount of risk.

CEO – The aforementioned strategies and countermeasures were executed and deployed successfully for our Bank. We exercised options to sell loans and swap positions, recording interest and gains totaling $4.5 million and delivered solid shareholder returns.

We ended 20XX achieving the following key financial performance metrics:

1) Highest profit in our peer group totaling $6.8 million.
2) Highest stock price in our peer group of $49.18.
3) We consistently returned capital to shareholders having the highest cumulative dividend payout of our peer group of $2.00 per share.
4) We ended up in a positive liquidity position.
5) We ended with a loan to deposit ratio of 103%.
6) We ended with a loan loss reserve coverage ratio of 1.2%.
7) We ended with the highest ROAA of .97%.
8) We ended with the 2nd highest ROAE of 11.37%.
9) Lastly, we ended with an efficiency ratio of 74% as I am a very generous CEO, and I paid out some ridiculous bonuses to all the staff!

I am sincerely thankful to my family and my team for affording me the opportunity to attend Pacific Coast Banking School. I could not have completed the program without all their continued support and encouragement throughout my three-year term. The subject matter, professors, written extension assignments, simulations, and group of highly talented financial professionals with whom I interacted with and befriended, all left an indelible impression upon me.

I looked forward to leveraging my accomplishments, experience, and added knowledge gained as a leader in the financial industry. Little did I know that everything I just learned at Pacific Coast Banking School would actually have me exiting my organization thanks to Tesla.

It was at the end of 2018 that I unknowingly decoded two key elements within Elon Musk's Secret Master Plans and unlocked the Multiplier. What Elon noted in his Master Plans, becoming energy positive, empowered as one's own utility, I unknowingly qualified and quantified into numbers. I did this by building a financial model to review with the family to reassure them that we made the correct financial decision in acquiring our Tesla's and Solar Energy System. Seeing the financial savings our family would realize, I figured why not build a model for everyone in the U.S. to use?

It took me two days to build after compiling gasoline and electricity metadata for every U.S. state. I ran multiple data simulations and was blown away. I went back and read Elon Musk's Secret Master Plans. His plans crystallized into numbers in my mind. The more I read and pondered, the more those secrets revealed themselves to me.

After building the Multiplier model for the entire U.S. I came to unlock a dynamic element revealed within the Multiplier: That the more Electric Vehicles one acquires, the more one saves due to the higher cost structure of gasoline relative to the nominal incremental amount of solar energy needed to eliminate both gasoline and electricity costs. I brought The Energy Matrix, Trinity, being Triple Hedged concepts to life building the Multiplier model, and just like that, I immediately hacked it! All I saw were numbers, multipliers, savings, and FREE! Talk about having the epiphany of a lifetime!

Once I unlocked the Multiplier model, it would lead me down the path of what I refer to as "The Tesla Challenge." It was nothing like my banking school challenge and is discussed in the next chapter. But for now, back to banking school.

It was also during this time that I finished my Pacific Coast Banking School report on Dynamic Leadership along with my Asset Liability, Interest Rate Risk, and Enterprise Risk Management Reports. Part of my research on leadership led me to Steve Jobs commencement speech delivered to Stanford's 2005 graduating class. Steve's speech resonated with me on how the collective experiences and challenges both in my life and career have connected the dots both past and present.

Steve Jobs noted, "It's impossible to connect the dots looking forward; you can only connect them backwards. You have to trust that the dots will somehow connect in your future. You have to trust in something, your gut, destiny, life, karma, whatever. Because believing that the dots will connect down the road will give you the confidence to follow your heart even if it leads you off the well-worn path, and that will make all of the difference.

You have to find what you love and that is as true for your work as it is for your lovers. Your work is going to fill a large part of your life and the only way to be truly satisfied is to do what you believe is great work. And the only way to do great work is to love what you do. If you haven't found it yet, keep looking. Don't settle. As with all matters of the heart, you'll know when you find it. And like any great relationship, it just gets better and better as the years roll on. So, keep looking. Don't settle."

Looking backward after completing Pacific Coast Banking School in 2019 provided me with much introspection about where I was at career-wise. It honed my analytical and leadership skills that much deeper with perspective however, I graduated conflicted as a leader. Not because of the leadership sessions I was engaged in at PCBS, but of the opportunities that lay ahead. Do I pursue a more prominent leadership role in the financial services industry? It's a fundamental question of leadership discussed at Pacific Coast Banking School to size up our leadership development. We were all asked, "What have we accomplished as our personal best as a leader?" It's a straightforward question. However, I was conflicted. Yes, I had done many good things cumulatively throughout my career but what was my personal best when it came to my profession? I always acted as a Trusted Financial Advisor to my colleagues, company, and clients. However, it was always on a relatively small and narrow scale making them millions or saving them millions. I needed to think bigger if I really wanted to make a real impact as a leader.

Part of my conflict (of my own accord) was layering in another dimension to the question of what have I accomplished as my personal best as a leader? The added layer? How was I also developing future leaders? Sure, I had worked on many significant projects leading my teams. From seeing through a whole enterprise systems conversion to burning the midnight oil to get regulatory filings done. I had also worked diligently to get others promoted on my Team. Nevertheless, I always felt destined and capable of doing much, much bigger things in my career. It made me realize I was too comfortable with what I was doing and where I was at. This is what bothered me. Everything I had accomplished during my career up to this point were what I believed to be all good and great things as a leader. But on a magnum opus level? Not yet. I have high expectations for myself and needed to step outside of my comfort zone.

We needed to share our leadership personal best with groups that we were separated into at the Banking School. I have always mentored and challenged my teams, but somehow, this was just me doing my normal job as a leader. I just didn't feel I had a career story that stood out with a real WOW! factor.

It turns out my personal best I spoke to my team about was a personal story and not a career based one. On a personal level, it was easy. It was about me being a leader to my two children, Garrett, and Kristen, and as a husband to my wife, Debbie. I wanted to surprise Debbie in a grand way for her 40th birthday. By grand, I mean pulling out all the stops – A Day to celebrate a lifetime!

17

A LIFE OF LOVE WITHOUT LIMITS AND A LIFE LIVED WITHOUT REGRETS™

"This may sound corny, but, love is the answer."

– Elon Musk

I look far and ahead into the future. I plan and review multiple scenarios to maximize the best outcomes both financially and personally. Planning for Debbie's surprise 40th birthday was right up my alley. I started planning, formulating, and sharing my ideas with the kids on what kind of surprise birthday I had in mind. It was a long journey that took two years to plan in advance, along with investing over 2,000+ hours (mainly by the kids). Why two years and 2,000+ hours? I wanted it not to be just a grand celebration, but for it to be a totally epic birthday surprise! Garrett, Kristen, and I wanted to surprise Debbie with a most memorable 40th birthday party at the Ritz-Carlton.

We first started our planning by partnering in the kid's

piano teacher and shared our ideas for Debbie's surprise party. I asked their piano teacher to select and teach the kids a special piece to which they would perform at the party. Their teacher was in total shock that we were planning for Debbie's party two years in advance and that it would be at the Ritz-Carlton. She understood how serious we were and the importance of this event. After a couple of weeks, she pulled together a couple of sheet music options. We agreed upon a genuinely challenging, but beautiful piano duet by Anton Arensky published in 1894. Waltz (Valse) 4 hands from 6 Pièces enfantines Opus 34/4. The sheet music was wicked crazy! It looked like hundreds of scattered ants due to the level and varying tempo of notes combined with being a duet piece. Granted, Garrett was only nine and Kristen was only seven at the time, with only one year of piano lessons behind them. Add in the fact that I have no piano experience whatsoever meant the kids were truly on their own in pursuing this. I thought they would never be able to learn something so extremely complicated. The piano teacher assured me she could teach them in two years as long as the kids remained committed to practice. A lot of what I envisioned for Debbie's party was riding on this major piano duet, all of which I did not know how was going to turn out two years into the future. One thing the kids and I understood is that we were all-in and were committed to seeing this through.

Every morning Debbie would leave for work between 5:45 am - 6:00 am. Each morning after she left, I woke the kids up to practice their duet for 30 – 45 minutes while I fixed their breakfast and got them ready for school. Without question, Garrett and Kristen would dedicate themselves secretly to practicing every morning like this for the next two years. The not so secret or funny thing is, that Debbie had grown highly frustrated over the kids and their piano lessons over these two years. After school the kids would practice their piano lessons. However, it was from sheet music they had already learned. They did not have the time to learn new material on top of the Arensky duet. Debbie kept questioning why the kids were playing the same piano pieces repeatedly over and over again. She commented that they mastered them, and it was time to move on to new material. Debbie was totally unaware, not realizing Garrett and Kristen were secretly learning a totally crazy and wicked duet behind the scenes for her birthday. Time and time again, she would approach me, saying the kids were not learning new material and that they had a horrible piano teacher. I would assure Debbie repeatedly that the kids are playing at their own pace and to give it more time. It was a challenging time over these next two years managing her expectations while the kids were diligently playing every morning secretly after she left for work.

It didn't stop with just the piano lessons. There was much, much more to plan for. I took the kids with me to meet with the Event Planner at the Ritz-Carlton for meal tasting options in order to finalize the dining menu. Even though I could have done this on my own, I wanted to involve and teach the kids every step of the way about what goes into planning a surprise party. The Event Planner was taken back at how far in advance we were planning the event, along with me bringing the kids from a learning standpoint. Typically wedding planned events are done one year in advance but to plan for a surprise birthday two years in advance was something she had never seen before. Also, she said having a nine and seven year old participate in the meal tasting options was cute and heartwarming on a whole different level.

When time permitted, the kids and I also worked together building an epic video slide show for Debbie. It was important for me to work together with the kids on capturing 40 years of her life and our life together as a family on video. We also needed the entire two years to pull the video together as we have experienced so much together as a family. Getting the music to tie into the pictures and video fluidly took the longest amount of time during the edit and rendering process. It gave us an opportunity to reflect on how special our family's journeys, experiences, and adventures have been. We celebrated every moment reliving these experiences together.

Fortunately, Garrett and Kristen were able to successfully master the Arensky duet in one and a half years. Their duet together was absolutely stunning. Their piano teacher could not have selected a more moving or emotional piece. But why stop there? We had another six months before the party. So, me being me, asked their piano teacher if she could also teach the kids new and separate solo pieces. Their piano teacher knew we were all-in going for the WOW! factor. Now, you think the kids would just about have me committed to an institution with the idea of now learning an additional piece? They didn't. They didn't complain. They didn't moan. They didn't say the duet was enough. Garrett and Kristen understood how truly happy and proud this would make their mom feel, so they were fully committed. I approached the teacher with the idea of letting the kids select their own solo pieces.

Garrett selected and learned to play *Colors of the Wind* and Kristen selected and learned *My Heart Will Go On*. This meant even more piano practice on top of staying synced with their Arensky duet. Needless to say, I had the easy job as Debbie's birthday approached closer and closer. My job was making sure the kids practiced piano, helping them with their birthday speeches, finishing up the video, and work on even more surprises I had planned: hiring a Polynesian dance troupe, booking a family trip to the Big Island of Hawaii, and buying her dream car, a Lexus LS model.

Debbie's BIG day had finally arrived. The magic hour was upon the kids and I. On the day of her 40th surprise party, Garrett who was now eleven and Kristen who was now nine, had a cousins sleep over the night before. This was so that Debbie's sisters could take the kids to the Ritz-Carlton early to make sure the venue was in order: tables, centerpieces and to set up for the piano, audio, and video.

I told Debbie that I was taking her to a posh restaurant for her birthday and that she needed to dress her best. She was magnificently stunning from head to toe! She was wearing a contouring black dress sprinkled with patterned faux diamonds. I was wearing a silver/light grey 3-piece suit. The restaurant I told her we were going to just happened to be on the same route as the Ritz-Carlton. This was so that she wouldn't suspect anything unusual during the commute to the restaurant.

The party plans were in motion and detour plans were set. On the drive there, we were chatting about the kids and what vacation destination we could plan for later in the year. As we drove to the stop sign where the Ritz-Carlton was, I made a detour into their parking. Debbie asked what I was doing. I told her I was thinking of changing our dining plans and wanted to check out the Ritz-Carlton since it was on the way. All of the Ritz-Carlton staff in the front lobby knew who I was and were expecting our appearance. My eyes met with the Event Planner who was awaiting our arrival. We both made eye contact and tipped our heads in acknowledgement that the surprise was all set. I walked Debbie to the open atrium courtyard where family and friends were waiting. They all greeted Debbie and I with a roaring "SURPRISEEEEEEEEEEEE!!!" It was all coming together as I had envisioned. Debbie and I made the rounds greeting everyone who were already enjoying cocktails and hors d'oeuvres in the open atrium courtyard.

Unfortunately, as time drew closer to heading into the ballroom Kristen had developed a stomach ache. She had become sooooooooooooo nervous and was almost in tears. Kristen was scared of playing the piano and of giving her speech in front of everyone. I kneeled and held both of her hands. When she told me how she was feeling I understood why she was so nervous. At nine years old, I didn't think of all the pressure she was under in doing something so big. I reassured her that Garrett would be right next to her the whole time. It would be just as they have practiced at 6:15a.m. every morning these last two years. Telling her that Mom and I would also be right there at the piano next to her would hopefully be the added courage she needed to see this through. I encouraged her that if she did this, it would show her mom what a big girl she had grown up to be. Her mom would be incredibly proud to see the level of devotion and determination she would demonstrate on this one day. I told her to take deep breaths and to relax. Kristen, "Today will come and go, but it will be a memory that will forever last a lifetime. Your mom will be so very proud of you. We have been planning this event for two years to show that our love to her is a love without limits. Let's make it an epic experience just like all the other experiences and memories we have accomplished together as a family." She was a little more at ease and nodded her head in agreement. I kept my fingers crossed that it just might be enough to take the edge off her nervousness.

After an hour in the courtyard, we adjourned into the ballroom. This wasn't just any ballroom. It was a ballroom with gold-guild vaulted ceilings, stained glass windows, and oversized Swarovski chandeliers. It gushed with beauty, resplendence, and grandeur. More than a century of historic charm and events unfolded in this ballroom and now we were going to be part of it.

Everyone took to their assigned seating in the ballroom. Now, the nerves started kicking in. We started the event with the kids and I each taking turns giving our speeches. We spoke of just how much Debbie meant to each of us. The kids thanked her of course for always being the perfect mom. I thanked her for being the perfect wife, for giving me this wonderful family, for challenging me to achieve more, and for always bringing out the best in me. Our lives together have been an absolutely fantastic journey together and we have come so very far in celebrating every moment of it. I talked about the not-so-funny story of the kid's piano lessons and how frustrated Debbie was at their lack of progression. She of course, had no idea that the kids had spent the last two years preparing for this one special day. The moment of truth was now upon the kids. Would Kristen's nerves get the better of her, or would she come through and deliver? It was a lot of pressure to ask of a nine year old who had never done something so big or so public before. I hoped with fingers crossed that she would find the strength and courage deep down inside to pull this off.

I asked Debbie to stand by the piano and motioned Garrett and Kristen to begin. As they began to play their duet, I joined Debbie by the piano holding her right hand. Their duet followed by their solo pieces could not have been more audibly and visually perfect. The kids totally and absolutely nailed it! They never missed a key, nor a beat, nor were off tempo being synced as a duet. Debbie was beyond moved and taken back at just how beautiful their duet and solos were. She now realized the magnitude of the kid's efforts over these last two years devoting over 2,000+ hours secretly practicing at 6 am every morning when she left for work. After all the horrible stories of how terrible their piano teacher was, it was this same piano teacher who was the one directly responsible for picking out the duet. It was through her efforts in working diligently with the kids over these last two years that had revealed the results of those efforts. All our nervousness washed away. The kids and I breathed a sigh of relief. Everything we had planned for, executed on, and delivered is what we had envisioned.

Their piano performances were then proceeded by the video we put together of Debbie's life and our lives together as a family. Capturing her life over these 40 years of her growing up and our relationship together as a family only furthered our love for her. I had spent months and hundreds of hours arranging and coordinating the song set list to accompany the video slideshow. It needed to be perfect, and it needed to be moving. It needed to capture celebrating 40 years of her life.

The venue location, the ballroom, the speeches, the piano duet and solos, and the slideshow video demonstrated our devoted efforts over these last two years. However, we weren't done yet. As our meals were being served everyone was dazzled with fun and entertainment by a phenomenal Polynesian Dance crew performance. The Polynesian dance troupe covered all dances from the Pacific: Hawaiian, Tahitian, and Fijian. At the end of the show performance, the main dancer presented Debbie with the keys to her brand-new Lexus LS that we arranged to be parked out front of the Ritz-Carlton with a large red bow. Oh yeah! But we still weren't done yet. The last and final surprise announcement that we made was that our family would be leaving for vacation the next day to the Big Island of Hawaii, staying at the Hilton Waikoloa Village. Two years of advance planning encompassing over 2000+ hours came together just as we had epically planned and envisioned it. We could not think of a better way to spend our family vacation but at the Hilton Waikoloa resort swimming with Dolphins and chilling poolside with some lava flow drinks in hand.

Everything turned out to have much more of an effect than the kids and I could have ever imagined. My mother told me afterwards when she was in the bathroom, there were more than half a dozen other females crying besides her talking about how beautiful everything came together from our speeches, the kid's piano pieces, and Debbie's video life story. Gathering and putting together the pictures of Debbie's life and our family's life together were one thing but sequencing them to the music soundtrack captured and encompassed the multidimensional aspects of our lives together even more emotionally than I could have ever envisioned.

Two years of extremely high-level planning, commitment, attention to detail, dedication, and execution to plan Debbie's surprise party came and went so quickly. Her 40th birthday celebration event was yet another life experience love and joy multiplier in demonstrating the amazing journey our family has been on. This was Debbie's special day, a day celebrating a lifetime. It was befitting as everything we worked on collectively over these last two years demonstrated our family's devotion to each other and how we have lived our lives – **A life of love without limits and a life lived without regrets™**. It is a life philosophy that is dear to our hearts and one that keeps our family high on life.

This was me as a leader, as a father, and as a husband leading my family to do our personal best and bring out the best in each other. Teaching and guiding Garrett and Kristen about my vision, planning, and execution, along with Debbie as my partner, would help to shape them as leaders throughout their lives and be instrumental in defining their own legacies.

Garrett would go on to become Varsity Team Captain of his high school's Water Polo Team and a Blackbelt in Nippon Kempo. Kristen would go on to become Varsity Team Captain of her high school's Drill Team. Both graduated from college and now work at Fortune 100 Companies. They are the owners of the other two Tesla Model 3s in our family. They both happily made the decision to trade in their ICE vehicles and took on all financial responsibilities associated with their new Tesla's. Debbie is having difficulties parting with her Lexus LS due to significant sentimental reasons. It was always her dream car, and it was a very special birthday gift to her. Yes, it's totally hypocritical of everything I discuss in this book, but in all honesty, we've put less than 200 miles on it over the last two years just to keep the spark plugs clean and not let the fuel go stale. We have been working remotely from home due to Covid, so we don't need to drive two cars. We drive the Tesla to both avoid fuel costs and do any more environmental harm. However, the time has now come. It is time to put the Lexus out to pasture. I just recently (and secretly) ordered Debbie a Model Y, our fourth Tesla. We will soon be an all-Tesla Family.

Now these next few sections are going to be total tangents, but hopefully I am going to bring it all together. Fast forward to 2020 through 2021. I used this time to enlist the kids input to help me build multiple iterations of the model to ensure the coordination and arrangements were simple and straightforward. They are both business major professionals and were schooled as little Quants by yours truly. It goes back to the old adage: Give a kid a fish and they eat for a day. Teach a kid multidimensional data dynamics, quantitative analysis, risk management, strategic planning, and leadership (sprinkle in Piano lessons to enhance spatial reasoning) and they will be enriched for a lifetime. Isn't that how the adage goes? Funny, I think I taught my kids a little something different.

Once completed, multiple iterations were filed with the U.S. Copyright Office to ensure intellectual property protection. Calculations and formulas cannot be copyrighted, of course. However, the visual illustrations, coordination, and arrangements can be.

There was one last solid piece of intellectual protection I needed. It gave rise to formulating and self-publishing this book. This strategy would dovetail in perfectly as I continued on with my Tesla Challenge. If I published a book, just maybe it will find its way into the hands of someone at Tesla. It needed to be big, bold, robust, and was precisely what I was looking for. I have never written a book before. However, since I had written so many reports at Pacific Coast

Banking School, this allowed me to leverage from this experience and incorporate what I wanted to qualitatively and demonstrate quantitatively. Because the kids and I had built multiple iterations of the Multiplier model, new concepts, phrases, and illustrations described in this book all came to me fluidly over a short period of time. It would also give rise to new time dynamics needed for research, development, and implementation of additional intellectual property protections.

Consequently, I couldn't enlist the help of professionals such as book publishers or editors due to all the intellectual property collectively formulated. Therefore, please excuse any grammar issues, repetitiveness, and how the look and feel of this self-published book came out.

Okay, we've talked in detail about the Multiplier and "X." There are more interesting facets to "X." X.com was co-founded by Elon Musk. In 2001, X.com rebranded as PayPal. In 2017, Elon Musk purchased back the rights to X.com. On July 7, 2017, Elon tweeted, "Thanks PayPal for allowing me to buy back X.com. No plans right now, but it has great sentimental value to me." If you type in X.com now, the only thing that shows up is "X" in the upper left-hand corner. SpaceX was formed in 2002 by Elon. In May 2020, Elon and Grimes welcomed their son, X Æ A-Xii. He is referred to as "X." X stands for the unknown variable. Æ represents Grimes elven spelling of Ai, which means love in Japanese, and AI also stands for artificial intelligence. A-Xii represents Lockheed's Skunk Works A-12 Mach 3 reconnaissance aircraft. Lockheed's A-12 was the 12th in a series of internal design efforts for "Archangel," the aircraft's internal code name.

"X" is subliminally revealed throughout all constructs of Elon's vision and companies. Elon has shown and discussed "X" many times. Yes, his Secret Master Plans imbue details of a much, much more visionary design and strategy. They seem simple on the surface but are intricately dynamic and multi-faceted in terms of their structure, execution, symbiotics, and overarching goal. The competition doesn't see it based on the directions they are taking. I discovered a single facet of "X" by unlocking a secret element within The Energy Matrix. This single facet allowed me to qualitatively write this book and demonstrate it quantitatively with the Multiplier model. More facets of "X" are there, and no, I'm not just talking about my Multiplier or his son X. As I dug deeper and deeper, I uncovered more facets of "X" within Elon's vision and strategy. I refer to them as "Multidimensional X-factors." They are there. You just have to be willing to open your eyes and jump down the rabbit hole without a parachute. How deep down the rabbit hole did I go? The multidimensional X-factors inspired my vision for the book cover art and are embedded into it.

Why does all this matter and where is this all going? The final image of Elon along with all the dynamic elements went hand in hand during the completion of this book. Like the Multiplier model, the cover image appears simple but is complex in its meaning and development. It was simple because it had to incorporate an image of Elon Musk. However, it was complex because a book cover image is two-dimensional. This book, along with the cover illustration design absolutely needed to be big, bold, robust, and more precise in the message being conveyed. I needed to transform a two-dimensional image of Elon and incorporate multidimensional dynamics much the way my mind processes things. The complex part? How do I integrate multidimensional facets of Elon and his vision in a two-dimensional form?

How complex is complex? Multiple layers of complexity, of course. The book cover art for Elon's image was inspired from Tesla's March 31, 2016, Model 3 unveiling. It's a relevant moment in time as this live-streamed webcast was the basis for ordering our first Tesla Model 3s. It also represents Tesla's first mass-produced consumer Electric Vehicle. The Tesla Model 3 was an all-in, make it or break it moment in history for Elon and Tesla. The Model 3 is a key defining invention and product directly responsible for changing the course of Humanity. It's looks, performance, safety, technology, and price shook the legacy automotive world to its core. The Model 3 is the single greatest catalyst for the legacy auto manufacturers triggering them to go all-in on Electric Vehicles. This event is relevant in time-stamping Elon's image to capture his vision and execution for consumer adoption of electric vehicles on a mass level.

The artistic thought encompasses our family's global travels from the Picasso Museum to Antoni Gaudí's La Sagrada Família in Barcelona, the National Gallery in London, the Duomo Di Milano in Milan, the Sistine Chapel in Vatican City, the Galleria dell'Accademia in Florence, the Musée du

Lourve, the Musée d'Orsay, and Musée Marmottan Monet in Paris, to the Cherry Blossoms of Philosopher's Path in Kyoto and Meguro River in Tokyo. All these adventures, experiences, and beauty ran through my mind producing a visual kaleidoscope of inspiration. Walking along Philosopher's Path in Kyoto in April 2019 with my family while being immersed in a Sakura storm gave me the final transcending vision I needed for the book cover art piece. I researched more world global art and finally settled on blending highly geometric, analytical, interlocking, and orphic cubism dynamics leveraging vivid Simultanism sensations from Robert Delaunay, *Simultaneous Windows on the City*, 1912 coupled with Jean Metzinger's, *Sunset No.1*, 1906 – wide and vivid color spectrums. I settled on geometric cubism using Wedha's art genre as it had to incorporate multidimensional facets of "X." I also incorporated SpaceX's Starman's view set as the backdrop because a vivid crystal blue Earth represents another dynamic element imbued within Elon's Secret Master Plans. All of this was also done with the determination, vision, and inspiration taken from elements of Debbie's 40th birthday party slide show...that complex. Sound complex? It gets even more complex.

The final art piece for the book cover is aptly named: *"Immortal Magnum Opus – Elon Musk Multidimensional Dynamic Facets Transcending Space and Time."* Magnum Opus represents the greatest achievement of an individual. Why Immortal Magnum Opus? The full details of this will be revealed in the self-titled chapter, *Elon's Immortal Magnum*

Opus. My vision was to capture and artistically illustrate the multidimensional dynamics and facets of Elon's vision, determination, and success in his drive for the greater good of Nature, Humanity, and Earth: An Immortal Magnum Opus that transcends space and time. Everything collectively represents multidimensional symbolisms as follows:

1) Elon's illustration was brought to life and was designed to be symbiotic to the multidimensional dynamic facets that came about deriving the Multiplier models and principles formulated in this book.

2) The multidimensional facets correspond to Elon's Secret Master Plans and how each element gives rise to his Immortal Magnum Opus.

3) The art's taut, geometric interlocking cubism dynamics also represent multidimensional facets to formulate Elon's facial shape delivered at the March 31, 2016, Model 3 unveiling. A date, invention, and engineering masterpiece that marks a precipice moment, and turning point in the future of Humankind – the first mass produced consumer Electric Vehicle that disrupted multiple industries and ideologies around the world.

4) The vivid color spectrum dynamics within the multidimensional facets that make up Elon's face also represent the collective colors of all Earth's living creatures - Nature, Humankind and Earth. How highly brilliant, beautifully interconnected, and interdependent they are upon each other for both survival and prosperity.

5) The geometric intersecting multidimensional facets give rise to multiple "X's" as a result of the Multiplier model, Elon's appreciation of "X" along with corresponding multidimensional X-Factors imbued within his Secret Master Plans.

6) Elon's eyes convey the intense passionate vision, determination, and success of his mindset. The blue color intensity of his eyes rightfully parallels Earth's crystal blue backdrop and represents his vision for a crystal blue planet Earth.

7) The vivid crystal blue beauty of the Earth represented in the background represents the overarching goal of Elon's Immortal Magnum Opus: scaling Trinity throughout the world for the collective advancement of Nature, Humanity, and Earth.

8) When taken in collectively, all the symbolisms of the multidimensional facets when acted upon by all of Humankind with the same passionate vision, determination, and success reveal a selfless love for Nature and Earth – A life of love without limits.

9) Yes, it sounds all very ingratiating towards Elon Musk. He is after all, a huge inspiration that gave rise to many elements formulated in this book. Nevertheless, it's also meant to convey and capture all the symbolisms of the beauty that awaits the future of Earth the sooner the world transitions to Trinity. All of Humankind needs this kind of inspiration, determination, vision, and resolve to see this through for Earth's true beauty to be realized.

So, the art cover illustration is just a little complex but

like with Debbie's party, I'm not done yet. As I mentioned, I jumped far down the rabbit hole without a parachute. Additional art pieces were developed and are still being developed. They might eventually be revealed along with the entire collections name but it's a secret for now. There are more multidimensional immersive elements that accompanies the art. It involves incorporating many of the audio songs that the kids and I used for Debbie's 40th birthday, plus others that are meant to bring all the multidimensional dynamic elements together. It's a song arrangement set list from Josh Groban along with incorporating My Immortal from Amy Lee of Evanescence that we used in Debbie's video slide show. These songs are vocal and instrumental power ballads that resonate love, loss, resolve, and triumph. The visual aspects of the art collection are designed to accompany the following song collections and are to be listened to in this precise order for the full inspirational effect:

1) Mi Mancherai – *Closer* by Josh Groban
2) Oceano – *Closer* by Josh Groban
3) Solo Por Ti – *Awake* by Josh Groban
4) When You Say You Love Me – *Closer* by Josh Groban
5) Awake – *Awake* by Josh Groban
6) My Immortal – *Fallen* by Evanescence
7) You Are Loved (Don't Give Up) – *Awake* by Josh Groban
8) Alla Luce Del Sole – *Josh Groban* by Josh Groban
9) Un Giorno Per Noi (Romeo e Giulietta) – *Awake* by Josh Groban

Why these specific songs? I listened to them while revisiting Debbie's video slideshow. This song set list and Debbie's video slideshow were instrumental in formulating thoughts and components of the artwork and many elements in this book.

Why did I end it with Un Giorno Per Noi, the love theme from Romeo and Juliet? Please make note of Un Giorno Per Noi. Look up the English translation on Lyrics.com after reading this book. No, not now. At the conclusion of the book. I know you are curious. However, the conclusion of this book will bring everything together.

The art collection is meant to be viewed in a dimly lit setting with a 4,000K color temperature down lighting effect focused on the art pieces. This is needed to bring out the intensity of the vivid color spectrum dynamics along with evoking emotions conveyed in deep thought and introspection of our lives to date and the type of future Humanity will create on Earth. (Note: Unfortunately, there is a color differential in the printed contents and book cover formatted version due to RGB versus CMKY coloring rendering issues for printed materials. I am developing an electronic book ("eBook") version which will render the color spectrums more accurately for the ebook contents and cover.)

If you are interested in a little sample, go ahead, and give it a try after finishing this book. Download the song set list. Then take the book cover and set it under lighting with a 4,000k color temperature. Set the room lighting tone to your liking. Pour yourself a glass of your favorite beverage. Get comfortable. Take a deep breath. Take a sip of your tasty drink. Start the audio playlist. Sit back, relax, and reflect on your life to date and the future you will make.

So, you see, the book cover while seemingly simple, is highly complex and animated in the thought and vision that went into producing it. The artwork along with the audio selections are to inspire and give rise that Humankind is capable of demonstrating and experiencing love and joy multipliers with Nature and Earth – A life of love without limits.

Now on to the website. I'm particular about the look and feel of certain things. When I designed and built the Energy Trinity website, it was no different. I have no experience in building or designing a website however, all the content and layout that I designed formulate the basis for Trinity. Every aspect of the website has an underlying meaning. Some assemblance of this is revealed at the end of the book. I've had my plate full, so for now, the website is on-going work in progress that will grow over time with added content.

Back to graduating from Pacific Coast Banking School. Now I needed to find some way in delivering my personal best as a leader. It has always been easy for me to identify opportunities strategically for-profitability positioning, cost saves, and risk mitigation for organizations benefiting the C-Suite, shareholders, and clients. But I had grown too comfortable and bored only because I had such a great and knowledgeable team. After graduating from PCBS however, I wasn't challenged so my work wasn't satisfying or rewarding.

I thought, "Well I might get picked up by a big bank and I might save or make them billions if I stuck to banking. However, the idea of saving consumers trillions and saving Earth is way more exciting!"

Interestingly, the future direction of my career would be further catalyzed by Steve Jobs and tested by Tesla. If I had never enrolled into PCBS, I would have never written about or immersed myself in Dynamic Leadership. I would have never researched Steve Job's commencement speech on connecting the dots. I would have never interacted with so many other talented Banking industry leaders. I would have never been afforded the opportunity to demonstrate my financial and strategic tactical skills in leading my team to 1st place. If I had never enrolled into Pacific Coast Banking School, nor took a deep dive analysis on Asset Liability, Interest Rate Risk and Enterprise Risk Management, nor written numerous analytical reports, nor acquired three Tesla's and a Solar Energy System, I most certainly would not have decoded two key aspects of Elon Musk's Secret Master Plans or developed the Multiplier model for the entire U.S., nor build the Energy Trinity website, nor write this book, nor come up with the ideas or coining the phrases, words, and concepts such as The Energy Matrix, Trinity, being Triple Hedged, Immortal Magnum Opus, nor invoke deep thought as to the design into the book cover art illustration. These were all new elements in connecting my life's dots that helped to further shape me and my life's direction.

Another key take away from my experience at PCBS during the Dynamic Leadership sessions was also about defining and leaving a legacy. We were asked, how do we want to be remembered at the end of our lives? What kind of legacy will we leave behind?

I never really thought about leadership on that level, and it was just what I needed. It was an indirect punch in the gut to me, but in a very good way. It brought my research on Steve Job's and his unfortunate passing into perspective. It stopped me cold in my leadership assessment tracks and made me ponder about what leadership I could demonstrate to cement my legacy as I continued forward on my career path. The humbling part I came to terms with when establishing my own legacy was tasking myself to write my own obituary. It wasn't required or asked by Pacific Coast Banking School to do. I put it upon myself to assess what I have accomplished in my life to date and what I still have left to accomplish. Sounds morbid to write one's own obituary but it was the right leadership perspective I needed at the right time in my career and life. It crystalized and brought new perspective on how I have lived my life to date, what I have accomplished, and what more I wish to accomplish and experience in life.

I needed to think bigger as a leader if I truly wanted to make a difference in the world. The bigger the impact of my actions, the more meaningful legacy I would be able to establish. Yes, how we choose to live before we die is something we have some degree of control over. Death is something we have no control over. It cannot be fooled. It cannot be bargained with. It cannot be bought. No one can escape it no matter how rich, how educated, or how famous one is. Death comes to us all. What we choose to do from the time we are born up until the moment we die makes all the difference. Our life experiences and actions give us the ability to demonstrate leadership in making a mark towards our own legacy in how we contributed to the betterment of our families, our friends, our colleagues, our organizations, our communities, Nature and of course, Earth.

This thought invoking introspection on leadership gave me just the perspective I needed in furthering my leadership skills. After building the financial model, I could have been complacent, passive, and smitten keeping the model to myself. I would take great pleasure knowing my families Tesla's and Solar would be free and that we are well positioned against future energy risks. As a leader however, it didn't resonate with me. Steve Job stated, "I want to build really good tools that I know in my gut and my heart will be valuable. Then you just stand back and get out of the way, and these things take on a life of their own." I needed to share my savings, my strategy, and my knowledge with everyone. I needed to build the Multiplier model for everyone in the United States and write this book to demonstrate how they can become energy independent – being energy positive, empowered as one's own utility. By doing so, consumers stand to recover a majority, if not all, of their Electric Vehicle and Solar energy costs and eliminate their carbon emissions. I needed to see this through as it's a win for consumers finances and an environmental win for Nature and Earth. What I am doing collectively and how consumers respond to it will help count towards my leadership and my legacy journey.

A key aspect of leadership and legacies can be summarized as making this world a better place than when we came into it. Hopefully, I too, will cement my legacy by decoding key elements of Elon Musk's Master Plans, building and unlocking the Multiplier model along with formulating all the ideas contained in this book. I did so to awaken consumers within The Energy Matrix in order to save them a fortune in addition to accelerate the world's transition towards global energy independence. I did so for the greater good and for the betterment of Nature, Humankind and Earth. That's what a good leader would do and should do.

It has not been without challenges. The biggest challenge being trying to bring all this information to Tesla's attention. My family and I would go on to call it, "The Tesla Challenge."

18

THE TESLA CHALLENGE

"When something is important enough, you do it even if the odds are not in your favor." – Elon Musk

It was around November 2018 that I originally created a financial model to demonstrate to Debbie that we had made the right decision to acquire our Tesla's along with installing a Solar Energy System. I did this as we were looking to acquire a third Tesla Model 3 for our family. The acquisition of three Tesla's and a Solar Energy System was a large financial outlay. I wanted to reassure Debbie that the financial benefits over the long run would outweigh the upfront costs.

I built the financial model in under half an hour after researching energy prices for both fuel and electricity. It is well known in the Solar Industry and those who have adopted solar as to what their solar investment payback and breakeven is. It can vary from 7 – 15 years, depending on the state a consumer is in and what their electricity consumption and rates are. My original analysis for my investment payback on just electricity for our solar energy system was 7 years. When I added the cost and energy consumption dynamics of our Tesla Model 3s by eliminating fuel costs, it reduced our solar payback to 4 years. This is the moment I unlocked the key and beauty of the Multiplier. I discovered that the more Electric Vehicles one acquires, the more one saves due to the higher cost structure of gasoline relative to the nominal incremental amount of solar energy needed to eliminate both gasoline and electricity costs. I triple-checked the model dynamics and was eagerly smitten in walking Debbie through all the numbers and the dimension aspects of the model. Walking her through all the numbers validated that we had absolutely made the right decision to acquire the Tesla Model 3s along with installing a Solar Energy System. A third deposit on our next Tesla for Kristen was on order. Oh joy!

The financial results were so transformative I figured if we could save this much money and generate an even faster payback, why not build it for the entire United States? Challenge accepted! I researched, compiled, and embedded energy and pricing metadata dynamics for both gasoline and electricity for every State in America along with every vehicle and model along with their average mpg in the model. It took me about two days to build it out. It was straightforward, clean, and mind-blowing when extrapolating for how much consumers could save in the future by eliminating both gasoline and electricity expense.

As of 2020, there were up to 100 million single-family homes and businesses in the U.S. that stand to benefit directly from Solar. These consumers also own and drive over 275 million registered ICE vehicles. I then extrapolated the future cost savings and the financial value impact (should all U.S. consumers transition to both Electric Vehicles and Solar Energy) is jaw-dropping to the tune of saving $100 trillion in future energy expense costs for both gasoline and electricity over a 40+ year time period.

Starting in late 2018 and early 2019, numerous multidimensional model layouts were filed with the U.S. Copyright Office. Numerous iterations were built and submitted to ensure that the intellectual data coordination, arrangements, and layouts could not be reimagined by those looking to capitalize for their own financial gain.

It was around this time that Tesla also had their vehicle referral program wherein after 5 referrals, a consumer would get a 2% discount off a Roadster and with each 5 additional referrals get sequential 2% discounts off. Certain YouTubers had been so successful in getting viewers to use their referral codes that they were able to earn a Roadster for free.

I figured with my newly created model showing consumers on YouTube how to get a free Tesla and how much money they could save with Solar, I too, would try my luck. I was not a YouTube content creator. After conferring with legal counsel, uploading the video on YouTube would afford additional intellectual protection on top of the data coordination, arrangements, and layout submissions. I created a new account and published a 36-minute video on YouTube going over multiple scenarios with different cars and states talking about the costs savings of Electric Vehicles paired with Solar, solar warranties and being energy positive.

I thought the video would be picked up and trending as word spread about how to get a Tesla for free with Solar. A free Tesla Roadster would surely be mine! Sadly, the video didn't escape YouTube's algorithm black hole as there were no video views. So much for putting in key words like, "TESLA", "FREE TESLA", "SOLAR", and "FREE SOLAR" in the title header and video tags. I have since switched the video from public to private and filed the video with the U.S. Copyright Office. Why? Because certain crafty Youtuber's skirt on the edge of other's published video content to rebrand as their own. These actions would not allow others the opportunity to leverage off specific content. Because of the Multiplier model's financial significance and intellectual property value in getting free Electric Vehicles and Solar, certain YouTuber's and Solar Company's might look to exploit changing aspects of my public video as their financial hustle. Should YouTuber's or Solar Company's discuss the combination of getting Electric Vehicles and Solar free with Solar they will most likely trigger a copyright violation strike.

So, what next? I compiled a list of 45 email addresses of many high-profile news agencies, business journalists, Solar Companies and Tesla. I sent them a link to the YouTube video with hopes of it being looked at. No one did. I don't blame them. If someone I didn't know sent me a video link, I too would simply delete the email. Not knowing what the link was to or if it was a virus is always a concern. My emails could have also gone to their spam inboxes.

What next? I needed to get the Multiplier model in front of Tesla, so I tried calling. The problem was it is impossible for any member of the public to get through to anyone in any department at Tesla at the corporate office. I get it. They just launched the Model 3 and every team member's focus was on ramping up sales and delivery logistics, among hundreds of other things going on at Tesla.

In late January 2019, I emailed Tesla's North American Press and Vulnerability Reporting group regarding whom I could speak to. They were the only emails I could find. I got no responses.

In February 2019, I created a twitter account and tweeted dozens of times to @ElonMusk. I also mailed paper hard copy examples of the Multiplier model to Elon Musk and to, now former Chairman, Brad Buss at their Fremont Corporate Office. Still no responses.

In May 2019, I drove to the Tesla Regional pickup center where we picked up our Tesla's to speak with a supervisor. It was during this time that the Model 3 deliveries were ramping up leaving me few staff members to speak with. I was able to catch one employee's attention. I pulled out my laptop and went over the model with him and he understood it. I thought, finally, someone gets it. It's just a matter of time before I show a supervisor and that supervisor continues escalating it within Tesla. He provided me with his phone number and said he was reaching out to his supervisor and to Elon's personal assistant. A week had lapsed with no response from the original Tesla employee. I texted him and was informed that his supervisor was too busy to meet with me and that Elon's personal assistant never responded to him. I understood how busy they were as the regional pickup center was absolutely jammed packed with owners waiting to take delivery of the new Tesla Model 3s. Still, I received no follow-up response.

On May 22, 2019, with multiple submitted copyright iterations of the model in place, I decided to demonstrate the model to a team of bankers encompassing an Executive team and other various staff. Nothing gets the attention of Bankers letting them know you are going to show them a multi-trillion dollar model. I cracked open the model and went through a couple of "What-if" scenarios quantifying the impact. They all got the model immediately and were floored at the financial results. I just gave them the highlights and did not reveal the Multiplier effect. That was a little secret I was keeping to myself. Why? Dollar signs $$$ lit up their eyeballs like a Vegas slot machine! The financial model was a total hit. They were blown away that it was built for the entire United States. It gave me further affirmation that it was understandable, straightforward, and easy to follow. Imagine if I had shown them the Multiplier effect coupled with the fact that the more Electric Vehicles one acquires, the more one saves due to the higher cost structure of gasoline relative to the nominal incremental amount of solar energy needed to eliminate both gasoline and electricity costs? They would want in on the action and try to capitalize on it. I have had far too many instances wherein others capitalized on my financial works, and I was not about to let this one get away from me.

I was totally energized and amped up about the positive results from the Multiplier model. I talked to Debbie and kids about the situation. I told them that because I had been unsuccessful in getting through to anyone at Tesla my next best option would be to go directly to their corporate office. They knew I was on a mission, and they fully supported my decision.

May 29, 2019 – I visited my cousin who lives in the Bay Area which was ideal for visiting Tesla's (now former) Fremont Corporate office location. I didn't want to show up late in the afternoon at Corporate because of time constraints in wanting to connect with the right individuals at Tesla. I opted to explore Palo Alto and grabbed dinner at Ramen Nagi. Decent, but nothing like Ichiran's in Tokyo.

I had already made up my mind earlier that I was going to camp out in my Tesla. I drove around and found a sweet spot in a swank residential neighborhood in Palo Alto. The Tesla blended in perfectly and no one suspected a thing. I put up the sunshade on the front windshield, dropped the back rear seats, rolled out my sleeping bag, pulled out my laptop, and stress tested the Model some more preparing for the big day tomorrow.

May 30, 2019 - I had a great night's sleep and was ready to visit Tesla Corporate after an early morning shower. I had made reservations at the Watercourse Way Day Spa. It was just the ticket I needed. I relaxed in the jetted spa and showered up in their Three Trillium water room before arriving at Corporate.

I rolled up at 3500 Deer Creek Road in Palo Alto however, I didn't see any guest parking. SAP Software Solutions was directly across the street, so I took my chances parking there and walked over to Tesla's lobby. I checked in at the reception desk noting that I did not have an appointment. I explained that I was a Tesla owner and shareholder and wished to speak to someone about a financial model I built that Tesla would greatly benefit from. The receptionist conferred with another associate and made a couple of calls while I sat down with my laptop. I was told by the receptionist that an individual named Martin XXXXX (last name withheld for confidentiality reasons) would meet with me and that they needed my I.D. to check me in. I provided them with my driver's license which they copied, and I also provided my business card showing that I was a senior financial officer.

I continued waiting for about 15-20 minutes when I was approached by a Tesla Employee whom I initially thought was Martin. It turns out I was approached by an undercover security member of their corporate office. He asked me about the purpose of my visit. I advised the security member that I had built a financial model that would help sell more of Tesla's products. I proceeded to tell him I did not have an appointment because I could not connect with anyone at Tesla. He advised, unknowing to me, that there was a security check-in booth that all guests needed to check-in at visitor parking. Because I parked across the street, I did not know there was visitor parking and that I needed to check-in. I just walked into their lobby from across the street. There was no special key or security badge needed to get into their lobby.

The security team member then asked for my I.D. I

handed him my driver's license and business card. I explained to him once again that I had built a financial model that would greatly benefit Tesla however, I could not get in touch with anyone at Tesla by phone, email, or in person. Because I could not get in touch with anyone, I decided to come directly to corporate in hopes to get anyone to look at the model. I advised him that I was also a Tesla Shareholder and owned multiple Tesla's. This did not seem to sit well with him as he continued to grill me about the purpose of my visit. I could see his demeanor and tone towards me was making the front receptionist uncomfortable. He then advised me I was to leave immediately. I told him that I was advised by the front desk that Martin XXXXX would be seeing me. The security guard advised me that that was not going to happen in an elevated tone. At this moment, another receptionist showed up. I could see both receptionists started feeling uncomfortable. I could sense they felt just awful at how I was being treated. Wow! It just dawned on me at that moment that I got the bait and switch. So much for seeing Martin. I felt like I had just committed a serious crime. I re-iterated to the Security member that that I was a Tesla shareholder and owned multiple Tesla's and the financial model I built would help Tesla sell more of its products. All I needed was to show it to anyone at Tesla who understands Finance. I even asked if he would like to look at it. He declined and asked me to leave immediately.

I got the message, but I wasn't quite done. There were employees who were exiting out of the main lobby. I quickly explained to a group of them that I built this model for Tesla and if I could just show it to them. I could see the anger in the Security Guards face, so it was a huge gamble. Needless to say, the group of Tesla employees brushed me off and kept walking towards the exit. Talk about feeling sheepish. I closed my laptop and made for the exit. The security guard who asked me to leave had two other security members follow me out to my car parked across the street making sure I had left the corporate campus.

Wow! This did not go as I had planned at all. I get that I did not have an appointment, but I could not get over the fact that I still could not speak to anyone at Tesla. Not by phone, not by email, not by social media, not in person, and now not with anyone at Corporate. Talk about striking out - Times 100!

I recalled that Elon hung out in the lobby of Netscape in the 90s hoping to talk to anyone about getting a job. He was too scared to talk to anyone, so he just left. I wasn't even looking for a job with Tesla. I was just trying to help Tesla sell even more of their products by trying to get anyone at corporate to listen to me. Instead, I was asked to leave. The security guards continued eyeing me in the SAP Software parking lot to make sure I would not be returning. I drove off around the corner and parked at Hewlett Packard on Page Mill Road digesting what just happened. I understand the need for security in this day and age, but wow, I was a Tesla shareholder and owned multiple Tesla's. I felt sheepish and ashamed.

I mustered up enough courage to call Debbie and let her

know what happened. She and the kids had been so supportive. They understood I was trying my best to help Tesla but was out of options. They believed in the model as did others that I have shown. I was beyond frustrated that I couldn't get just one person at Tesla to look at it.

It was a long drive home with much introspection. I am always used to providing successful winning financial solutions. Because of the size and scale of the model that would benefit Tesla, Consumers, and Earth, I felt that all my efforts thus far were not only futile but also foolish. I felt defeated in a way I have never felt before.

Alas, I would not give up just yet. Fortunately, the receptionist provided me with Martin's email address before the whole security event unfolding. I followed up and wrote to him to explain why I went to Corporate without an appointment:

Re: Appointment

From: ▮▮▮▮▮▮▮▮▮▮▮▮▮▮▮▮▮▮▮▮▮

To: mv▮▮▮@tesla.com

Date: Thursday, May 30, 2019, 05:57 PM PDT

Martin

I'm so sorry it just dawned on me that I inadvertently called you by Mike on my last email....my apologies. Thanks!

▮▮▮▮▮▮▮

> On May 30, 2019, at 2:58 PM, ▮▮▮▮▮▮▮▮▮▮▮▮▮▮▮▮▮▮▮▮▮▮▮   wrote:
>
> Hi Mike
>
> My apologies for dropping by unexpectedly there at corporate without an appointment today.
>
> I figured I would stop by to see if I could speak to anyone at Tesla before returning back ▮▮▮▮▮▮▮▮▮▮. I was   in NorCal visiting my cousin over the Memorial Day weekend and figured since I had my laptop I would try to connect with someone at Corporate. ▮▮▮▮▮▮▮▮▮
>
> I am a Tesla Shareholder and also own 2 Tesla Model 3's....absolutely love them!
>
> I built a dynamic financial model that pretty much destroys gasoline burning vehicles. It is perfect for Teslas website as it allows a consumer to see what the financial cost savings are in going with a Tesla and complimenting it with a properly sized solar panel system.
>
> The thing is I cannot connect with anyone   at Corporate to show the model to. I tried calling many multiple times and cannot get in touch with a live person at Tesla.
>
> The model doesn't have to be in shown person. It can simply be as easy as a webex session.
>
> I really want Tesla to succeed in its mission in accelerating the world to sustainable energy. This model I built will do just that for Tesla for increased sales for both vehicles and solar. I believe it is exactly what Tesla needs to put all the negative news as of late to

```
rest.
>
> If you are not the right individual to connect with this I would appreciate if you could
assist in connecting me with the right individual.  As a Tesla shareholder and Tesla owner,
I would very much appreciate it.  I truly believe just as Tesla is revolutionizing the
automotive/energy industries this model I built demonstrates to consumers why they
should financially
>
> Thank you again for your time
>
>
>
>
```

Needless to say, I never heard back from Martin on my emails. This still did not deter me. I would have one additional chance – Tesla's Annual Shareholders meeting on June 11, 2019.

June 11, 2019 – My family is Tesla shareholders. When we received our shareholder proxies for the annual meeting, I responded that I would attend in person. It was being held at the Computer History Museum located at 1401 N. Shoreline Blvd. Mountain View, CA. Despite my failed efforts just a little over a week before at Corporate, I didn't need to tell my family that I would still be going. They knew I was on a mission to get the Multiplier model in front of Tesla. I just needed to connect with the right Tesla employee at the shareholder's meeting.

I flew into San Jose Airport and arrived at the Computer History Museum at 10:30 a.m. This was well in advance of the 2:30 p.m. meeting start time. I arrived to a most gorgeous sight - a trio of red Tesla's yet to be released: The Semi, the Roadster, and Model Y. A big shout out to Frans and the design team! The ability to incorporate beauty while balancing both form and function in a Zen harmonized package is truly an artistic triumph! The entire product suite lineup is gorgeous, beautiful, and dazzling!

I wanted to get to the meeting early to better my odds in connecting with anyone at Tesla. I had my laptop in my backpack ready to go and fully juiced up. By noon, a line was starting to form at the front entrance to the meeting. I lined up with other shareholders and it just so happened that the fellow behind me was an Accountant by profession. A former Big 4 Ernst & Young alumni who left to work for a Bay Area tech firm. In our conversation, I told him I built a model for Tesla and was here to get it in front of anyone who would listen to me. He inquired about the model. I agreed this would be a perfect opportunity to show someone outside of my network who understands numbers. Being an accountant, he was familiar with data models. I walked him through the process and within a few minutes he got it. Not only did he get it, but he was also floored. He wanted to take video and pictures of the model, but I did not allow him to. There was also a lady behind him listening in on our conversation. She was a Tesla shareholder vacationing from Germany and decided to attend the shareholders meeting. I then walked her through the model, and she totally got it. She indicated that I must get this model to the German Automakers. I told her I had to get it in front of Tesla first and had no intentions of sharing with other automakers or solar companies at this time. She begged to take a picture of the model. I declined. Tesla is best positioned to capitalize on it and would be at a competitive disadvantage if I shared it with others. I felt reassured that if I can walk people with whom I just met through the model, then it would be a total slam dunk if I could just get 5 minutes with any Tesla employee.

The doors officially opened around 1 p.m. We went through security, checked-in, and made it to the auditorium. I was surprised to see how small the auditorium was and that there only appeared to be about 300 seats. I quickly grabbed a seat and scoped out other fellow shareholders.

The meeting was off to a little late start, but Elon eventually made it on stage. He gave everyone an update as to what Tesla has accomplished along with current challenges it was undergoing and the future direction of Tesla. There were a lot of great stats and specs on Tesla's growth and on-going strategy. Then the moment I was waiting for - shareholders were invited to ask questions. The mic was already set up just one row behind me but there was a mad scramble to the mic. Just like that, about 6-7 people got in line before I did. Elon fielded questions from both the right and left side of the auditorium. All I needed was to get my chance at the mic. My question was simple. I built a financial model that Tesla would benefit from that will boost sales of all its product lines. Who at Tesla could I get to just listen or do a Webex session with me for 5 minutes? As luck would have it, they stopped further shareholder questions with just one more person in front of me. Whiskey Tango Foxtrot!!! I wasn't ready to give up.

It just so happened that after the meeting was over Tesla's Chairperson, Robyn Denholm, was standing in front, center row floor. I recognized her and knew of her background as a former CFO. If anyone would understand my finance gibberish, she would be the right person at Tesla to speak to.

I walked up to Robyn, introduced myself, shook her hand, and handed her my business card. I proceeded to explain to her that I was a senior-level finance professional and developed a financial model for Tesla. Just as I was getting to the meat of the model, in steps a Tesla Employee. Seriously!

The Tesla employee who interrupted us was Kamran Mumtaz, (now former) Tesla's Director of Global Communications. It turns out Robyn was needed for a special board meeting that was being held adjacent to the auditorium after the shareholder's meeting. She stepped away from our conversation and joined up with Larry Ellison. I could not believe it! Now I had to start all over about the financial model with Kamran. I went on to explain to him the financial Model I built for Tesla and that it was crucial that I get to speak with Robyn about it. The only thing he could do is provide me with his email contact information. Kamran indicated I could follow up with him via email about getting in touch with her.

Wow, this could not be happening! She, along with Larry and other Board members proceeded straight into a private board meeting. I thought maybe I would just try to re-connect with Robyn after the meeting. Unfortunately, no one was allowed to loiter around, so everyone eventually had to exit the Museum. I would not get to speak to the one person who was most likely the best qualified to run the model by. This was happening all over again like some sinister episode of the Twilight Zone.

I was running out of options. My last chance would be to wait outside of the Museum and get the attention for any departing Tesla Employee. They were easy to recognize with their Tesla Badges/Passes with a lanyard around their neck. Geeze, I sound like a stalker. Finally, an Asian woman who was a Tesla employee happened to be walking towards my direction. She looked just about my age. I was hopeful that just maybe she could spare 5 minutes of her time.

I approached her and handed her my business card. I told her that I flew here just to speak with anyone at Tesla regarding a financial model I built that would bolster sales of their entire product line. I told her that I was not a salesperson and was just a regular Tesla owner and Tesla shareholder. I went on to elaborate about the model briefly and that if I could just show her the model on my laptop. Yet, she stopped and interrupted me from saying anything more. It turned out, she worked in Tesla's legal department. She advised me that she would not be able to look at the model from a legal standpoint. Spoken like a true lawyer. Seriously!!! She would not provide me with her business card but promised me she would have someone follow up with me once she consults with her team. Yes, I was officially in the Twilight Zone.

I flew home only to tell my family of the luck that would once again come my way. Once again, I felt defeated and deflated. I totally get that everyone is busy at Tesla. They were scaling production and deliveries. It was a make-or-break time for them, so I continued to give them the benefit of the doubt.

The week had passed by and still no word from anyone at Tesla or it's legal department. Fortunately, I had Kamran's contact details, so I emailed him on June 14th to follow up:

From:
Date: Friday, June 14, 2019 at 10:58 AM
To: Kamran Mumtaz
Subject: Hi

Hi Kamran,

I hope all is well.

We met at the end of the shareholders meeting and Robyn indicated you would be able to provide me with her email contact details. It's in regards to a Dynamic Financial Model build for Tesla's website that we were discussing before she needed to head into the BOD meeting. Thank you again for your help.

I was completely shocked! I heard back from Kamran the next business day:

From: Kamran Mumtaz
Sent: Monday, June 17, 2019 11:35 AM
To:
Subject: Re: Hi

Hi

Thanks for reaching out, and hope you enjoyed the weekend. Would you be able to send any additional details about what you had discussed? That way I have all the relevant information, and can share it with the appropriate team.

Thanks again, and look forward to hearing from you.

Best,

Kamran

I wrote back to him so he could put me in charge with the right individual or department at Tesla:

From:
Sent: Tuesday, June 18, 2019 9:36 AM
To: 'Kamran Mumtaz'
Subject: RE: Hi

Hi Kamran,

I had a great weekend. I hope you did too. I apologize for the long email however, I need to elaborated on the additional details.

Yes, I was speaking to Robyn regarding a dynamic financial model I created specifically for Tesla. As you know, Robyn is a former CFO so I figured she was the best choice to discuss the model with at the shareholders meeting since we both have backgrounds in Finance. However, as a fellow corporate officer, I understand Robyn is the Chairperson so I can fully appreciate this may need to be shared with an alternate team member for consideration/evaluation. The dynamic financial model I built is specifically designed to demonstrate to consumers the long-term financial benefits of acquiring an Electric Vehicle and complementing it with a solar production system. The cost savings from a financial consumer standpoint are substantial and are fully illustrated in the model. The financial savings provide perspective to consumers to encourage potential buyers of electric vehicles that solar is the best accessory option from a financial standpoint.

My family and I are Tesla shareholders and currently have two Tesla Model 3's VIN's # 8056 and 0010. We currently have another Model 3 in process for delivery by this quarter end RN# 6810. We are a firm believer in Tesla's vision in accelerating the worlds transition to sustainable energy. Fortunately, Tesla is currently the world's only company to offer end to end sustainable energy products from its electric vehicles, battery storage and solar panel systems. Because of this, Tesla is the only one to benefit from this dynamic financial model.

Before I took delivery of our two Model 3's last year, I researched, spec'd out, and installed solar on my home.

I spec'd out my system to make sure I had sufficient solar production to fully offset both my household electrical and electric vehicle consumption needs. Since I have a background I Finance I built a model to determine what my earn back or break even was on my initial cost outlay...easy enough. Because I have a background in Finance and data modeling, I took it to the next level. I then layered in how much in fuel transportation costs I was spending towards my (thankfully good riddance) gas burning cars. We effectively hedged our household energy and transportation costs going forward for the next 35 – 40+ years the moment we purchased our Tesla's and complemented the purchases with a properly sized solar panel system. We no longer have to pay the electricity company or big oil companies for our electricity needs or fuel transportation costs for the next 35-40+ years. We are no longer subject to the costly effects of inflation over the next 35-40+ years. I quantified these energy costs in the dynamic model.

Just like with any car, just because the warranty expires this does not mean the car does not work anymore. It just means certain materials are no longer covered for repair/replacement however, the car will continue to operate for many many more years. My wife's current Lexus came with a 4 year/50,000 mile warranty. It will be sad to see it go but it's a gas burner. Its inefficient and will eventually be replaced with by a 4th Tesla vehicle in our family. ☺ Solar production is much the same with warranties. Solar panels will continue to produce energy past their warranty however, performance production gradually diminishes (solar degradation) over time, hence a 35+ year useful life.

Because I have a background in data modeling and finance, I took the model to the ultimate level. I built it for the U.S. for every consumer to use. It forward projects both electrical and gasoline costs in every state of the U.S. out for the next 35+ years. It also captures every make and model by manufacturer for sale in the U.S. to demonstrate what those forward projected costs will be. It is a dynamic financial model that allows a consumer to input what state they reside in, what the cost per kilowatt is per state, what their average household electricity bill is coupled with the make and model vehicle they are driving along with how many miles they commute in order to provide prospective on what the cumulative costs of their electrical and transportation fuel costs will be. Once a consumer models the information and sees the how expensive projected future energy and transportation costs are they will be more incentivized to purchase a Tesla and compliment it with a properly sized solar panel system (who wouldn't want their own re-fueling station for the next 35+ years!) The model I built captures this both qualitatively and quantitatively.

All I need is 10-15 minutes where I can do a Webex session to show why it could be a game changer for Tesla. I believe the model will increase sales for Tesla for both electric vehicles and solar thereby increasing much needed cash flow for Tesla. I further believe the model will help accelerate Tesla's vision in a mass push towards sustainable energy. It is something no one on Wall Street has figured out yet. ☺

Thank you for putting me in touch with the team member on this. I truly appreciate it as a Tesla owner and as a Tesla shareholder.

Kamran was true to his word. He did forward my email to another individual within Tesla. It turns out the person who emailed me back was Martin. Yes, that same Martin who I was supposed to see at the corporate office back on May 30th before being removed by Tesla's Security and who didn't return my emails.

Fortunately, I was able to exchange a couple of emails with Martin regarding the model. I essentially just sent him the same thing that was already communicated to Kamran. However, I never heard back from him after this last email:

From: Martin
Sent: Thursday, June 27, 2019 2:16 PM
To:
Subject: RE: Hi

Hi

Sorry for coming back to you this late. I think this is a great idea, and definitely something we could embed to our website when we ask people to buy solar panels.

Let me first check with our Energy team to see if something along those lines is planned and potentially, how could we make it super easy for everyone to understand.

We'll reach out soon to have a look at your model through Webex.

All the best,

Martin

I never heard back from Martin. I never heard back from the Energy Team to do a Webex of the model.

I totally get it. I am not faulting Martin for not following up. Legal could have also advised Martin or the Energy Team to close the door on this matter. The thing is, I was a multiple Tesla car owner and a shareholder, so the entire experience was kind of mind-blowing not getting anyone at corporate to at least look at the model.

Maybe Tesla had something already planned at that time? Or maybe they didn't see the value of such a model? That was June 2019. Fast forward 3 years later and still nothing. Now I can say with certainty, that Tesla has not considered such a model because no such financial model exists on their website.

So, this was my Tesla Challenge to date. My twilight zone attempts at trying to get through to anyone at Tesla so they could sell even more of their products.

The aforementioned experiences didn't rest well with me however, I had to return my focus back to graduating from Pacific Coast Banking Schools 2019 session. I needed to put the Multiplier model on the back burner.

COVID-19 broke out in early 2020 and well, everyone can agree 2020 and 2021 were some absolutely crazy times. There was no way I would be able to get in touch with anyone at Tesla. Understandably so, given the global pandemic crisis.

Sadly during 2021, it was painful to watch Elon go through a shareholder lawsuit brought on by his acquisition of SolarCity. If only I was able to get through to anyone at Tesla. I firmly believe a demonstration of my Multiplier model would have ended the trial before it even started. The SolarCity acquisition will flip from being perceived as an extremely poor investment decision by investors, Wall Street, and the media to one of the most brilliant financial decisions ever made by Elon.

The idea of Solar isn't new to Elon. He formulated this plan for Solar way back in 2006, as noted in the first paragraph of his Secret Master Plans. He waited for just the right time to execute strategically. Solar was to fit in financially later, after the development and success of the Model S. As Elon noted in Master Plan Part Deux, *"Starting a car company is idiotic and starting and electric car company is idiocy squared."* If he stated he was forming an Electric Vehicle, Solar, and Battery Energy Storage company all at once, he would have been viewed as absolutely insane!   It would have been viewed as too much investment capital, human capital, vendor supply chain logistics, infrastructure, and synergies to vertically integrate all at once. Elon, leading Tesla, was on point every step of the way. It's all about vision, planning, execution, and timing. All these energy dynamic plans dovetailed in alignment with his Secret Master Plans. They were all contingent upon product execution encompassing manufacturing, sales, timing, and cash flow. There wasn't a single day where Elon didn't bring his business game face and strategy plans in building out Tesla.

Fast forward to today, 2022, 16 years after his 2006 Secret Master Plan and 6 years after his 2016 Master Plan, Part Deux and still, no one sees it. It further illustrates how poorly investors, the media, Wall Street, EV pure plays, and the Legacy Auto Industry understand about the significance of how integral Solar Energy is to Tesla's business model and the execution of Elon's vision.   The financial and environmental synergies realized through the combination of Electric Vehicles and Solar cannot be understated in augmenting Tesla's business model.   It's something no Investment Management Firm, Wall Street, Venture Capital Firm, Solar or Automobile company has come to realize. Incorporating Solar Energy and Battery Storage into Tesla's ecosystem as a result of Elon Musk's strategy unlocks tremendous hidden intrinsic value in Tesla.

Sure, I could have pushed down the model on the Internet, but would it be picked up in Google's search algorithm and get any traction? I didn't have much faith in YouTube's blackhole algorithm from my YouTube video experience. I needed to develop this book and it needed to be big, bold, robust, and more precise to continue on with my Tesla Challenge. I also needed something more concrete in the form of intellectual property right protection if I was going to publish the model. I envisioned shady YouTuber's, Solar, and Electric Vehicle Companies trying to exploit the model's intellectual value by taking credit for it and then trying to monetize it for their own financial gains.

One might ask, why didn't I just go to Tesla's competitors like Rivian, Lucid, Ford, GM, or VW with the Multiplier model? Why didn't I go to hedge funds like Softbank, Apollo, Sequoia, or Kleiner Perkins? Why didn't I go to Investment Bank's like Goldman Sachs or Morgan Stanley? Why also didn't I go to Amazon, Apple, Google, or Microsoft with the Multiplier model? If I did, it would have put Tesla at a strategic disadvantage. Besides, how would they intrinsically value the Multiplier? A model that saves global consumers hundreds of trillions of dollars globally while generating hundreds of billions of dollars in annual sales revenue for decades to come? A model that changes the Electric Vehicle race and was built to accelerate the world's transition to energy independence? A model that decodes two key fundamental elements within Elon's Secret Master

Plans? A model that unlocked "X", a multiplier dynamic that demonstrates that the more Electric Vehicles one acquires, the more one saves due to the higher cost structure of gasoline relative to the nominal incremental amount of solar energy needed to eliminate both gasoline and electricity costs? A model that changes both dynamics of transportation and energy forever?

One thing is for certain, I absolutely did not want the Multiplier model or the intellectual property rights of this book to fall into the hands of foreign countries. Tesla is an American company born and bred, as am I. Tesla has their S3XY Solar energy dynamics. I decoded key elements to unlock those energy dynamics. They are both American inventions by design and intellect through and through.

Sure, I could have gone to someone sketchy like a Slugworth in *Charlie and the Chocolate Factory* and sold Tesla out. I could have turned over the Multiplier model, but I didn't. I could have turned over this book and the ideas for The Energy Matrix, Trinity, being Triple Hedged, and Elon's Immortal Magnum Opus but I didn't. If I did, these competitors or firms would have at some point also figured out a few of my TESLAMAXSFAST strategy plays. TESLAMAXSFAST is an investment theme I formulated that further accelerates the world's transition to energy independence. If I sold out Tesla, these companies, and firms most likely would have loaded up the dump truck on holdings and brokered M&A deals or taken them private for a re-spun IPO. Doing so would give competitors or firms a strategic advantage over Tesla. It's also why I spent so much effort trying to get anyone to listen to me at Tesla. Despite going through the frustrations of The Tesla Challenge, I did not and do not want this to happen to Elon Musk or Tesla.

Elon and Tesla are the underdogs and continue to remain so by the competition, Wall Street, and the Media. Elon's spirit is true to the history of American invention and capitalism. His spirit and achievements directly correspond to the wealth he has amassed. Is he entitled to it? 100% absolutely without question he is! He assumed 100% of all the risks, and therefore, is entitled to 100% of all the rewards! His blood, sweat, and tears got him to where he is today. How bold a pursuit to challenge global legacy auto manufacturing and energy industries built from over a century of development and production? Couple this with his efforts in re-igniting space exploration. His vision and interpretation of what an automobile, transportation, energy, sustainable energy, artificial intelligence, and space exploration should be, has taken multiple global industries by storm through the formation and growth of Tesla, SpaceX, The Boring Company, Neuralink, and OpenAI. He has shown grit, determination, and unbending will to see his vision through in growing Tesla globally while creating over 100,000 jobs. His forward-looking vision, plans, and accomplishments has positioned Tesla as the world leader in accelerating the world forward towards consumer energy independence with its Energy Trinity S3XY Solar energy ecosystem. Elon dedicated crazy work hours going through production hell, sleeping on the factory floor, further validating his work ethos as a leader and the spirit of his determination and valiant efforts. He is a testament and example that with great effort and determination comes great success, and when that success is done right, comes great wealth. So yes, that which one sows, so too shall they reap. No one should question Elon's wealth. Not for a second. He has earned every bit of it. Never forget he assumed all the risks and invested his capital in growing Tesla, SpaceX, The Boring Company, Neuralink and OpenAI in developing them where they stand today. He is 100% absolutely entitled to all the rewards and accolades that come from those efforts.

Final thoughts: You can't put a price tag on a visionary leader whose S3XY Solar energy ecosystem will save global consumers hundreds of trillions by avoiding future gasoline and electricity energy costs while saving Earth, Nature, and the future of Humankind through the elimination of carbon dioxide emissions. Couple this with opening Tesla's patents to the world and efforts to advance and address future transportation needs and efficiencies gained with Full Self Driving and The Boring Company. Combine this with global internet access through the deployment of SpaceX's Starlink satellite constellation network along with pursuing advancements by Neuralink in Artificial Intelligence. Try placing a value on all of that, much less debating it? You can't, and I wouldn't. You can't place an intrinsic value on advancing Humankind while simultaneously saving Nature and Earth for future generations.

It's no secret that Elon published his Secret Master Plans for the entire world to see. He's been waiting for someone to decode primary key fundamentals of his plan and reveal The Energy Matrix. I hope I did them justice. Despite my multiple efforts in not being able to connect with anyone at Tesla privately, I want to believe that those measures along with the publication and ideas formulated in this book, the Multiplier model, my TESLAMAXSFAST strategy, and my book cover illustration were done so with the sincerest of intent. That collectively being, to position and elevate Tesla for even greater success.

After all these years, the family and I still think that it would have been awesome for Tesla to use, "What's your S3XY Solar Multiplier?" as a cool new marketing catchphrase. The same can be said about our TESLAMAXSFAST strategy. At least we think they are wicked cool in helping to further solidify and accelerate Elon's mission. We think he would have liked them both to increase sales and help broaden Tesla's strategic position further away from the competition.

Elon's quote reflects how I have pursued my life in going after any challenge when he stated, *"Persistence is very important. You should not give up unless you are forced to give up."* Now that I reflect back on the whole experience, I'm actually very fortunate and grateful to have gone through The Tesla Challenge. Had I not, I wouldn't have written this book, built the Multiplier for online purposes, come up with the book cover artwork, or built the Energy Trinity website. I'm not sure if this book will ever find its way into anyone over at Tesla, so for now, my Tesla Challenge continues.

So shines a good deed in a weary world.

19

TESLAMAXSFAST™

"I'd rather be optimistic and wrong than pessimistic and right."
– Elon Musk

I believe most investors have heard of Jim Cramer's coined FAANG publicly traded stocks acronym (FACEBOOK, AMAZON, APPLE, NETFLIX, and GOOGLE). Writing this book and building the Multiplier model has provided me with a broad, long-term view of what will play out these next few decades.

I also formulated and coined my own investment theme and acronym of companies that I believe are well-positioned to capitalize on the explosive and exponential growth opportunities in the new Trinity era. **TESLAMAXSFAST™** is an acronym I coined for the following publicly traded companies and ETF (Exchange Traded Fund):

T - Tesla	M - Maxeon	F - First Solar
E - Enphase	A - Alcoa	A - Apple
L - Lithium America's Corp.	X - U.S. Steel	S - Steel Dynamics
S - SunPower	S - SolarEdge	T - TAN ETF
A - AlbeMarle		

Yes, TESLAMAXSFAST is spelled all together and not

separated. This was done to create another unique and fanciful new trademark name. This was also done to avoid infringing on Tesla's name trademark.

I believe the aforementioned companies and ETF identified are best positioned strategically to scale the Electric Vehicle and Solar Industry revolution over the long term. It will take significant time and resources to scale exponentially, hence a long-term view. Additionally, I believe they are best positioned in making significant long-term strides in reducing global carbon emissions, making Earth healthier for all of Nature, and Humanity for many future generations. These Companies are also well suited for Investment Management Companies and Portfolio Managers to load up on from an Environmental, Social, and Governance (ESG) perspective.

Why did I publish them in this book and not keep them secret? If I'm getting the word out, I want all consumers to benefit and be enriched from it. Also, had I not done so, a few savvy Wall Street players would have figured them out sooner or later. Maybe not all of them, but a good handful of them. Don't expect the TESLAMAXSFAST composition to last long from a strategies standpoint.

One might ask, why didn't I just apply for a job with Tesla three years ago to get all of this in front of Tesla? I had far too many instances in my career where others have gotten promoted and rewarded for my ideas and work and I wasn't going to let it happen with this one. Besides, I can imagine how the interview would have gone. HR: "Please tell me, what skills do you have that Tesla would benefit the most from?" Me: "Well, I decoded two key aspects within Elon Musk's Secret Master Plans: Being energy positive, empowering oneself as their own utility. I built multiple financial data models for the entire United States by combining consumer electrical consumption, electric vehicles electricity consumption offset with solar energy generation using energy sector metadata utilizing proprietary algorithms which model dynamic "what-if" scenarios to demonstrate various projected future financial money saving multiplier outcomes compared against gasoline vehicles, gasoline fuel charges, and utility electrical costs. In doing so, I unlocked "X", an element within both of Elon's Master plans, a Multiplier – That the more Electric Vehicles one acquires, the more one saves due to the higher cost structure of gasoline relative to the nominal incremental amount of solar energy needed to eliminate both gasoline and electricity costs. It delivers a triple fatal blow to ICE vehicles, the Oil Industry, and the Electrical Utilities. Global consumers using the model stand to save $300+ trillion over the next 40+ years while Tesla stands to make hundreds of billions in sales for decades to come!" HR: "Well, okay. We will be in touch. Thank you." (Rolling their eyes and thinking in the back of their mind: What a complete nut job! We got ourselves a real loony here! NNNNNNNNNNNNEXT!!!)

As part of my never-ending Tesla Challenge, I will be sending a copy of this book directly to Elon and Zach Kirkhorn, Tesla's CFO. My only hope is that this book gets past their personal assistants and into their hands before the other market players do. They are the busiest people on

Earth at this moment so I wouldn't be surprised if it doesn't make it into their hands. Martin is out of the question. I struck out too many times with him.

Now for the high-level overviews of the TESLAMAXSFAST companies:

TESLA (TSLA) is the only Company leading the world's future for both transportation and energy independence. They are the world's only fully vertically integrated Company whose complementary suite of product solutions benefits consumers on a global scale which in turn benefits Earth on a global scale the most via the Trinity framework. As such, Tesla's energy ecosystem combination of Electric Vehicles, Solar, and Battery Energy Storage solutions makes it the only Company best positioned to exponentially capitalize as hundreds of millions of consumers around the world transition towards new electric vehicles and energy independence. Elon and Tesla consistently challenge the status quo on every level encompassing engineering, software, supply chains, and manufacturing. This spirit and execution have disrupted multiple industries across the globe and will continue to do so for the foreseeable future. If you have ever wondered why Tesla commands such a high multiple relative to future earnings, these are just a few of the known dynamics. Remember, Solar has been missed by everyone from Mainstreet to Wall Street, legacy automakers, and current EV only pure plays. Solar unlocks tremendous hidden value in Tesla. Regardless, my guess is after this book and the Multiplier debuts, many Wall Street analysts will suddenly be hiking their price targets on Tesla. The same can be said about the TESLAMAXSFAST investment theme. Surprise, Surprise. I just wonder what they might be basing their price hikes on? ;-)

ENPHASE (ENPH) sells home energy solutions for the solar photovoltaic industry in the United States and internationally. With over 400 patents and pending patent applications filed around the world, Enphase Energy is a global leader in patent filings covering renewable energy technology. Its patented microinverter technology for solar panels has revolutionized the solar industry with their technology. Microinverters convert direct current (DC) solar panel energy into an alternating current (AC) which is the form of electricity used around the world. As of December 31, 2021, Enphase has shipped more than 42 million microinverters, and approximately 1.9 million Enphase residential and commercial systems have been deployed in more than 130 countries.

SUNPOWER (SPWR) was founded in 1985 by Stanford Electrical Engineering Professor Richard Swanson. With over 35 years of dedicated solar experience, SunPower is the only U.S. based solar company that's been around longer than their 25-year product warranty with their end-to-end fully integrated solar energy solution. They currently have exclusive access to the world's highest efficiency solar panels featuring SunPower Maxeon cell technology. SunPower has received more than 1,000 patents for its solar innovations.

SunPower has a distribution and product platform to scale exponentially due to forthcoming robust solar demand.

LITHIUM AMERICA'S CORP. (LAC) a white lithium diamond in the rough...a really rough, rough one at that. Lithium America's Corp is actually a Canadian-based resource company focused on the advancement of two significant lithium projects: the Cauchari-Olaroz project ("CauchariOlaroz"), located in Jujuy Province, Argentina, and the Thacker Pass project ("Thacker Pass"), located in northwestern Nevada, USA. Geologists estimate that Nevada could potentially hold 25% of the world's lithium supply. However, the Company to date, has not generated significant revenues and has relied on equity and other financings to fund operations. The underlying values of exploration and evaluation assets, property, plant and equipment and the investment in Cauchari-Olaroz project are dependent on the existence of economically recoverable reserves, maintaining title and beneficial interest in the properties, and the ability of the Company to obtain the necessary financing to complete permitting and development, and to attain future profitable operations. Will Nevada yield white lithium diamonds?

ALBEMARLE (ALB) is currently the world's largest lithium producer for Electric Vehicle batteries. Lithium, the main materials component in EV batteries, is not considered to be rare. However, Albemarle mines a significant amount of lithium from two world-class mines. One is located in the Salar de Atacama (Chile), and the other one in Clayton Valley near Silver Peak, Nevada (USA).

MAXEON (MAXN) is a spin-off from SunPower and headquartered in Singapore. Maxeon is well-positioned for the world's global solar stage. Their patented Maxeon solar technology was built from over 35 years of boundary-pushing solar DNA, making them a global leader in solar innovation and efficiency. Maxeon Solar Technologies designs, manufactures, and sells advanced SunPower-branded solar panels to customers in more than 100 countries worldwide through a global network of more than 1,200 sales and installation partners.

ALCOA (AA) specializes in aluminum and is the world's sixth-largest producer of aluminum. A tremendous amount of aluminum will be needed by hundreds of millions of consumers seeking solar energy systems which use aluminum in solar panel racking and framing. Additionally, aluminum is needed in a lot of Tesla's as they are made with a blend of aluminum and high-strength steel.

U.S. STEEL (X) Wait where's the X? U.S. Steel's stock ticker is "X", and it is the 8th largest producer of steel in the world. Their trademarked XG3 steel is the most advanced of Advanced High Strength Steels (AHSS) in the automotive market today. It incorporates an optimum strength-to-weight ratio for improved safety and fuel efficiency; superior formability for winning style and lower per-vehicle costs. It's designed specifically to provide automakers with the most cost-effective material to design safer and lighter vehicles. A lot of this steel and technology will be needed due to exponential growth in Electric Vehicles.

SOLAREDGE (SDGE) is based in Israel. Their SolarEdge DC optimized inverter system maximizes power generation at the individual PV module-level while lowering the cost of energy produced by the solar PV system. Since beginning commercial shipments in 2010, SolarEdge has shipped over 29.5 Gigawatt ("GW") of its DC optimized inverter systems and its products have been installed in solar PV systems in 133 countries.

FIRST SOLAR (FSLR) was founded in 1999 and is a leading global provider of comprehensive photovoltaic (PV) solar systems using advanced module and thin-film solar panel technology. Their customers consist primarily of utilities, independent power producers, commercial, and industrial companies. Strongly financially positioned, First Solar has the platform to scale exponentially due to future robust solar demand.

APPLE (APPL) It's no secret that Apple has been working on Project Titan comprised of automotive projects that could ultimately lead to an Apple Car. Many years have lapsed about a definitive partnership or joint venture. From a strategies standpoint, however, they would not partner with any of the legacy ICE auto manufacturers from an ESG perspective. Also, it would not be cost-effective for them to build an Apple Car in China and have them ship them to America, not to mention all the mega fossil-fuel pollution that would result from shipping from overseas. Moreover, both strategies run counter to their environmental goals.

Will Apple go full-blown Electric Vehicle, Solar, and Battery Energy Storage going toe to toe with Tesla right here in America? I don't believe so. It would be too capital intensive along with significant manufacturing time build out constraints. It would take up to a decade to see anything of measurable scale. What about partnering up with the new pure EV players like Rivian or Polestar? If Apple digs deep enough, its due diligence team most likely would come to the same conclusion I did. Too much risk exposure. Would Apple use its position to make a cash and stock offer for Tesla? The sheer size of their balance sheet, cash holdings, supply chain network, engineers, and global consumer reach could further augment their consumer end products from an automotive play. It was heavily speculated back in 2018/2019 that Apple would be a good suitor however, Tesla found its footing and emerged from production hell victorious. If a tender offer was presented, I would be opposed to such a transaction. There is still much of Elon's Secret Master Plans to execute on. If Apple steps in now, it could derail many future revenue multiplier channels imbued within those plans. Tesla would most likely adopt a wicked poison pill strategy to fend off such an offer.

Steel Dynamics (STLD) is the 3rd largest producer of

carbon steel in the United States. They are also one of the largest metal recyclers with a broad, diversified product lineup. There are currently 275+ million ICE vehicles in the U.S. that will be of zero value which will need to be melted down and re-purposed into new Tesla's. The unofficial numbers for Tesla's CYBERTRUCK orders range anywhere from 1 million to 3 million. That's a good amount of steel needed. Tesla's Texas Gigafactory is well-positioned to source metals from Steel Dynamics. It's Stinton Steel Mill is also located in Texas. The brilliance of the Cybertruck cannot be underestimated. Yes, looks are subjective around the Cybertruck however, there is oh so much more beyond its looks. Strategically, the Cybertruck will go down in the annals of automotive, business, and finance books as another brilliant play by Elon. The automotive world and Wall Street doesn't see it, but I see it.

TAN (TAN) Invesco's Solar Exchange Traded Fund – An exchange-traded fund with diversified portfolio holdings concentrated in the Solar Industry Energy Sector. Its current top portfolio holdings are Enphase Energy and SolarEdge Technologies. I believe the portfolio manager/team will be repositioning the portfolio holdings should this book find its way to them.

One may be wondering how come there are no ICE auto manufacturers that made the list? More interestingly, how come none of the new EV only players didn't make the list? Surely, one of them is destined to ride Tesla's coattails to fame, consumer, and financial glory, right? None of them pencil out this decade strategically. Tesla stands alone and will do so for quite some time. We will examine why in the next chapter and why any news of Apple dipping its toes into Energy Trinity would be met with muted reactions.

Remember I said it might not be beneficial to pay for your Tesla's and Solar Energy System in full even if you have the cash and financial means to do so? Consider for example, if you purchased a Tesla for $50,000 and another $30,000 for their Energy Solar System back in 2018. Assume you had the cash available to pay everything in full and did so. You will take satisfaction knowing you have no loans to pay back and your investments are free and clear. However, let's say alternatively you chose to put $30,000 as a down payment for both the Tesla and Solar Energy System and carried loans for $50,000. You alternatively used that $50,000 available in cash to purchase Tesla stock back in April 2018. Had you done so at $20 a split adjusted share and still held on to it today, it would be worth over $750,000 - more than enough to pay for the Tesla and the Solar Energy System multiple times over. So, paying cash and in full is not always a good thing, at least in this example. Every individual's financial position and risk tolerance is unique to their situation. This is by no means investment advice or recommendations. Investing involves risk, including loss of principal. Past performance does not guarantee or is predictive of future results. Any historical returns, expected returns, or probability projections may not reflect actual future performance. I do not provide tax advice and do not represent in any manner that the outcomes described herein will result in any particular tax benefit or financial gains. Prospective investors should consult with a certified financial, tax and/or legal adviser before making any investment decision.

Will Investment Management Companies, Hedge Funds, and Wall Street Analysts agree with my TESLAMAXSFAST picks and long-term view? Will they incorporate leveraging some, if not all, of the foregoing investment plays into their strategies, riding them to riches? As I look ahead strategically, I wonder if Cathie Wood of ARK INVEST will create a new active ARK ETF based on the aforementioned. If this book finds its way to her, I believe ARK EARTH ETF (ARKE) would be befitting of such a thematic fund name and ETF ticker. Saving Earth through disruptive innovation has its humanitarian and environmental merits.

Yes, my family and I are positioned in my TESLAMAXSFAST strategy. Our portfolio is heavily weighted in Tesla, Solar, Inverters, and Lithium plays. We are really light on the others. Sure, I could bore readers with some quant analytics in this book. However, numbers and strategies are always dynamic and never static. I see data, risk, and opportunities through a different lens. What I say in this book will be different a year from now as will it 5 years from now.

Let us continue on to see why no current Electric Vehicles or legacy ICE auto Companies make the TESLAMAXSFAST cut and why I believe Tesla has no competition.

20

TESLA HAS NO COMPETITION

"The U.S. automotive industry has been selling cars the same way for over 100 years, and there are many laws in place to govern exactly how that is to be accomplished."
– Elon Musk

Good competition is always healthy. It is needed to drive perpetual performance, efficiencies, and innovation. Whether the auto industry chooses to acknowledge it or not, they are directly participating in fulfilling Elon's vision of accelerating the world towards energy independence. How so? They have all adopted measures to completely phase out all ICE vehicles in the future. The legacy automotive world knows their ICE vehicles are obsolete. What the entire automotive world doesn't realize yet is that they will be directly responsible for plaid accelerating the world towards energy independence. It's one of the overarching goals of Elon's Secret Master plans. I will explain how this will come to be.

It's interesting to see news headlines regarding the perceived onslaught of competition against Tesla. Yes, there are dozens and dozens of new Electric Vehicles that are rolling out and are in the queue. However, that doesn't mean they are competition. Articles, publications, and YouTube videos about current EVs and forthcoming EVs that are "Tesla Killers" are misguided. I believe Tesla has absolutely no competition right now from a global standpoint. Will that change in the future? Sure, it will. But not this decade. If this were a race, Tesla has not only circled the racetrack, but it also continues to accelerate away from the competition. The competition believes they actually left the starting line. However, they will soon realize they need to stop, put it in reverse, re-strategize, re-execute, and then maybe attempt to creep slowly out onto the track. Allow me to explain my thoughts.

Electric Vehicle pure plays like LUCID, RIVIAN, NIO, BYD, XPENG, FISKER, FARADAY, and POLESTAR seek sustainability in pursuit of being ecofriendly. Their business models don't stack up against Tesla's vertically integrated Energy Trinity ecosystem. As a result, I believe they missed the mark and did their investors a disservice due to being strategically shortsighted. They might err and say that solar energy was in their strategic plans to begin with or a long-term plan of theirs. If they did, it would have been baked into their SEC S-1 filings for those that are publicly traded before they went IPO for an even larger subscription offering to account for the billions needed for the Solar and Battery Storage solutions buildout. BYD has been around since 2003. Lucid has been around since 2007. Rivian since 2009. NIO since 2014. Faraday since 2014. XPENG since 2015. Fisker since

2016. Polestar since 2017. Plenty of time has lapsed since their formations if they were to pursue Solar and Battery Energy Storage. As of September 2022, none of these EV pure plays have any plans to incorporate Solar Energy into their product solutions. Not now, not at their formation, not during their IPO. They don't yet understand the synergies Solar Energy offers relative to their own Electric Vehicle product offerings.

Will they opt to pursue the vertically integrated Energy Trinity model in the name of going vibrant green for investor and public optics? I believe they all need to. Those who do will have to raise more cash in exchange for control of some ownership or in the form of issuing stock or go to the debt markets by issuing debt. These measures will negatively dilute shareholder value or weigh negatively against earnings or both. It will be a painful learning lesson to their founders, C-SUITE/Board, shareholders, and venture capitalists in their rush to ride Tesla's coattails. Pursuing a merger and acquisition for Solar and Battery Energy Storage will be costly, time-consuming, and run into integration issues as will starting from ground zero. Pursuing solar partnerships provides little in the form of optics with no vertically integrated energy ecosystem benefit and a shortsightedness of a much-needed long-term strategy.

The U.S. represents the biggest Solar market opportunity globally. This is due to the tremendous solar irradiance it receives coupled with up to 100 million business and housing units with sufficient roofing available for deployment and installation that would support over 275+

million registered ICE vehicles that will all be replaced with Electric Vehicles. Tesla is the only pure play poised to capitalize on its vertically integrated S3XY Solar energy ecosystem. Its financial success will only be hampered by its ability to scale, source materials, and ability to fend off the many, many bad actors in the forthcoming energy revolution.

The publication of this book and the Multiplier model could transform the automotive industry into an era known as THE DECADE THE ICE AUTOMOTIVE WORLD STOOD STILL. The world's ICE automotive production future is now at an even more elevated risk of uncertainty. This has the potential to be on that scale. Am I fully aware of the implications and magnitude this will have globally, and the different industries affected? Yes, I am. Most rational and good-intentioned people understand that time is of the essence when it comes to saving Humanity, Nature, and Earth. Elon Musk certainly does. He started on this journey, against all odds, almost two decades ago.

Can Humanity, Nature, and Earth wait for legacy ICE auto manufacturers to phase out their vehicles, releasing billions of pounds of carbon emissions into the atmosphere these next few decades? How about your consumer finances? Could you afford to enrich the Monopolies by tens or hundreds of thousands of dollars more over the next few decades? Should I have buried everything not revealing The Energy Matrix, Trinity, the Multiplier model, the idea of being Triple Hedged, and let global consumers continue feasting on black pills, day after day, month after month, year after year? It could be argued that this would have allowed for a more orderly transition to Electric Vehicles. However, the world can't wait. Also, I believe you too would agree that neither can your future finances. Yes, it was weighed on the merits of the risks versus the rewards. It boiled down to save Humanity, Nature, Earth, and your future finances? Or continue to enrich the Monopolies and legacy ICE dinosaurs who mocked Elon and Tesla and chose comfort, laziness, and profit over innovation and solutions towards transitioning the world to sustainable energy? What would you do, let alone tell future generations if you didn't?

Let's explore some of the ICE legacy leadership comments on Tesla and their view on Electric Vehicles:

Bob Lutz, former President of Chrysler, former Vice Chairman of General Motors, and former EVP of Ford was quoted saying in May 2020, *"Tesla is doomed. They have no unique technology. All they have are some good-looking electric cars that sell below cost. With continued lack of profitability, the stock is going to tank. The next capital raise is going to be difficult. Everybody else has to sell a certain number of electrics. Not because there's huge customer demand for them out there. But they have to do it to satisfy government regulations. They will sell their electric vehicles at a loss, but they'll recoup it on the conventional vehicles whose sell is now assured because they did what they had to do on electric vehicles. So, the current economic model for all electric vehicles for all manufacturers is sacrificial pricing to get the needed volume to recoup the losses on your conventional vehicles. But here is the problem with Tesla. They have no conventional cars. They can't make money. Every single vehicle they sell, they are selling at a massive loss. The Model 3 isn't going to save them."*

2 years later and Mr. Lutz is pretty much 100% wrong on everything.

Akio Toyoda, Toyota Global President was quoted saying in December 2020, *"Electric Vehicles are overhyped. In a country such as Japan that gets most of its electricity from burning coal and natural gas, EVs don't help the environment. The more EVs we build, the worse carbon dioxide gets."*

This next section is devoted to all the climate change or global warming deniers and those who think Electric Vehicles are just as dirty or even more dirty than ICE Vehicles. It's sane, sensible, and simple. I will do it in about 3 pages. All things being equal, when you compare the lifetime emissions emitted from combustion vehicles stacked directly against the lifetime emissions from Electric Vehicles, it's a no brainer. For example, let's take the Earth down to a microscopic level. Now, envision the Earth is your home garage and the floor, walls, door, and roof represent the encapsulating atmosphere. I'm not suggesting this due to obvious safety reasons. But just envision for a moment, if one turns on a combustion engine versus one who turns on an electric vehicle in that enclosed garage? Which one do you think will emit more deadly CO_2 emissions over an extended period of time? Yeah, I know it's an absolute horrific thought. Now, multiply this by almost 1.5 billion polluting combustion emitting vehicles around the world over a 100+ year time period. Do you get the picture on a macroscopic level on what the impact is to Earth's atmosphere? This isn't just an absolute horrific thought, it's an absolute horrific tragedy, because it's real! I believe most sane people would choose the Electric Vehicle over the ICE vehicle 100 out of 100 times.

Regarding the EV manufacturing buildout creating more pollution, it too, is a no brainer. Combustion vehicles engines and transmissions have hundreds of parts (engine block, pistons, piston rings, cylinder heads, connecting rods, crankshafts, cams, camshafts, cam gears valves, valve covers, spark plugs, coil packs, engine control module, flywheel, starter, gaskets, alternator, belts, radiator, radiator fans, water pump, air intakes, power steering fluid and pump, thermostat, gas tank, fuel filter, fuel injectors, oil, oil pan, oil pump, oil filter, exhaust manifolds, catalytic converter, muffler, transmission, transmission fluid, and the related hundreds of screws, washers, and bolts that hold everything together...etc.) which are built in hundreds of factories that go into final assembly. Electric Vehicles require just dozens of parts that go towards their drivetrains and transmission. Tesla's Internal Permanent Magnet Synchronous Reluctance Motor is an engineering masterpiece designed around maximum efficiency, maximum power, maximum strength centered around minimal parts and minimal weight. So, let's see, hundreds of factories churning out hundreds of parts for ICE vehicles that require transporting and shipping those hundreds of parts globally. Compare that to an Electric Vehicle with dozens of parts made by dozens of factories that require transporting and shipping those dozens of parts globally. It really is no comparison. ICE vehicles emit substantially more emissions due to the production and transportation logistics involved in their manufacturing.

Lastly, let's talk about pollution from mining lithium versus mining oil. All things being equal, oil must be pumped, refined into gasoline, processed, stored, and shipped. The same can be said about most of the manufacturing aspects in processing lithium into batteries, so I will say it's more or less a push in terms of the energy intensity involved. The key difference are the end products and energy expended of gasoline versus a lithium battery. Gasoline instantly ignites, combusts, and then turns into deadly CO_2 along with other bad by products. Its energy is immediately expended. A lithium battery, however, can store, expend, re-store, and re-expend energy for hundreds of thousands of miles without any toxic emissions. Again, I am not suggesting this due to obvious safety reasons but just envision for a moment, you are in a fully enclosed camping tent for an extended period of time. There are two types of lighting options. One is a kerosene oil lamp and the other is a rechargeable battery-operated flashlight. Which one would you choose? Yeah, again it's an absolute horrific thought. Now, multiply this by almost 1.5 trillion gallons of gasoline a year consumed by global ICE vehicles and the resulting deadly CO_2 emissions released into Earth's atmosphere. I believe most sane people would choose the rechargeable battery over combustible fuel 100 out of 100 times. Once again, do you get the picture on a macroscopic level on what the impact is to Earth's atmosphere? Once again, this isn't just an absolute horrific thought, it's an absolute horrific tragedy, because it's real! Yes, it's repetitive, but I just wanted to really ingrain those images into your mind. Yes, Electric Vehicles emit far less pollution. It really is that sane, sensible, and simple.

One last parting thought. This is where Tesla gets it right and differentiates itself from all competitors. Tesla incorporates Solar Energy systems on all their manufacturing facilities as a hedge against emissions. They walk the talk and deploy capital where it needs to go. Just downright badass.

Jack Hollis, Toyota North America EVP of Sales was quoted saying in August 2022, *"It's unlikely mass adoption of Battery Electric Vehicles will develop as fast as environmentalists, the U.S. government, and most of the U.S. auto industry seem to expect. That's because, fundamentally, consumer demand just isn't sufficient. I don't think the market is ready. I don't think the infrastructure is ready. And even if you were ready to purchase one, and if you could afford it... they're still too high."*

Can someone in Toyota's sales department provide Mr. Hollis with Tesla's CAGR (compound annual growth rates) sales figures and its CAGR of global Supercharger network over these last 5 years to get an understanding of just what fundamental demand and infrastructure ready looks like?

Jim Farley, Ford CEO was quoted saying in August 2022, *"We are really on a mission at Ford to lead an electric and digital revolution for many, not few. And I have to say the shining light for us at Ford is this beautiful Lightning made right down the road in Dearborn, right here in the state of Michigan, already the leader of all EV pickup trucks in our industry in the United States. Take that Elon Musk!"*

Through July 2022 year-to-date, Ford sold 4,469 F-150 Lightning's compared to 564,743 Tesla Model S, 3, X, and Ys sold globally through June 2022. Yes, the Cybertruck hasn't debuted yet. Nevertheless, Ford will soon understand why Elon referred to the Cybertruck as Tesla's Magnum Opus and why it waited to debut the Cybertruck to incorporate its new 4680 battery architecture complemented with IDRA's NEO 9000 GigaPress.

Mary Barra, GM CEO was quoted in May 2022, *"We have said that by mid-decade we will be selling more EV's in this country more than anyone else. Including Tesla."*
2025 is fast approaching. Through July 2022 year-to-date, GM sold only 2,633 Bolt EV's in the U.S. Based on these sales figures I believe most would consider Ms. Barra's statement to be without merit.

Former Mercedes CEO Dieter Zetsche was quoted saying in September 2016, *"We can set a new target for ourselves and that is equally to be the leader in electric premium vehicles as well. Latest, by 2025 and this includes not only our current competitors but new entries as well including Teslu."*
Through June 2022 year-to-date, Mercedes sold 4,048 of its Electric EQS EV sedans in the U.S. Like with GM, 2025 is fast approaching and based on these sales figures I believe most would consider Mr. Zetsche's statement to be without merit.

We are at the precipice wherein consumers will now be

awakened and empowered that they are no longer subject to the Energy Dependent Monopoly Model. Consumers in the United States will start a movement transitioning to energy independence at an exponential rate. It will propagate globally. Consumers will stop purchasing ICE vehicles. Consumers will dump their ICE vehicles in mass for Tesla's Trinity S3XY Solar suite of products. As a result, look for the entire auto industry to follow in their footsteps, just as they have followed Tesla in their pursuit of Electric Vehicles.

Elon and Tesla directly upended and continue to upend the entire automotive world. Legacy ICE auto manufacturer's EV short-term and long-term strategies are highly questionable. Their business models have been obliterated by Tesla. Extremely painful choices and decisions will need to be made. Sure, they could acquire or partner with Solar Companies to give the appearance they are going vibrant green. Unfortunately, pursuing such a strategy will most likely further accelerate the demise of their profitable ICE vehicle business models. It's like a drug dealer who attempts to transition to sell ice cream. They are still building/selling ICE vehicles that pollute Earth while pursuing Electric Vehicles and Solar to upsell a vibrant green public image strategy? Consumers would see through this and opt for Trinity, leaving ICE sales at a standstill. Consumers will ask themselves, "Why buy an obsolete ICE automobile that is of zero financial value and continue to enrich the Monopolies?" Let us examine a few more challenges legacy automotive manufacturers face on a high level.

GM, Ford, Stellantis, Toyota, Honda, Subaru, Mazda, Kia,

Hyundai, VW, BMW, and Mercedes will certainly realign their business models pursuing Trinity for optics. The dilemma? It will be a cascading domino effect on a global scale that will most likely accelerate the demise of their profitable ICE product lines that much quicker. Legacy ICE auto manufacturers will rush to put out marginal, lower cost EVs to gain market share while balancing financial, manufacturing, and supply chain damage control. Now throw in Solar and Battery Storage development and integration into the mix.

For now, the real challenge to the legacy ICE automotive world will be tasked with managing two parallel business models. One for Electric Vehicles and one for legacy ICE vehicles, not for years, but for decades to come. Let us examine just a few of the issues:

1) Legacy ICE automobile companies have thousands of traditional brick-and-mortar franchise dealerships globally. What you see is what you pay with Tesla. We put down a deposit online, uploaded pictures of our ICE vehicles for trade-in value towards our Tesla's and uploaded our auto insurance and loan details in about 5 minutes. We also took delivery and picked up our Tesla's in a record 5 minutes. This is what an efficient and positive customer experience and engagement is. Haggling over an ICE vehicle has always been just downright ridiculous. A quality consumer product sells itself. Why have a middle person game the numbers in a final sale over many hours of back-and-forth bargaining, leading to

consumer frustration and ire? ICE manufacturers will need to run parallel platforms: an EV online platform and an ICE in-person haggle platform. Franchise auto dealers have significant financial and logistical pain coming their way as legacy ICE companies look to scale EV sales directly to consumers. How will Franchise's respond to getting by-passed and not getting a cut of the action? Dealer pricing markups are totally at risk of elimination. Consumers will balk at price markups over Tesla's what you see is what you pay platform. Auto manufacturers risk losing "a customer for a lifetime" should dealerships add price markups to their Electric Vehicle's and make consumers go through the whole sales sham ordeal.

2) New Vendors and Supply Chains. All new EV Battery and Solar vendor and supply chain relationships must be forged and sourced. Tesla leads the way with its vendors and supply chains. It's forged key relationships in key demographic locations globally. ICE auto manufacturers have a huge undertaking in sourcing new EV and Solar vendors. Look at the snafu GM had with LG's batteries destroying its Chevy Bolt Electric Vehicle platform. Other battery EV makers have yet to be fully tested. It's an edge Tesla has – proven engineering and battery performance with integrated software optimization along with almost 30 billion miles driven collectively by global Tesla owners. Once up and running with new vendors, they will need to run parallel supply chains: One on the new EV side and one on the ICE legacy side. They will still need to maintain the servicing on the outgoing ICE

vehicles soon headed for recycling and EV re-purposing. Servicing will still persist on 275+ million ICE vehicles in just the U.S. alone, and a total of almost 1.5 billion ICE vehicles globally for many decades to come. It will take significant time and resources as they get scrapped, recycled, and turned into EVs.

3) EV's are a whole different engineering class level than from an ICE mechanical manufacturing standpoint. Ask your ICE mechanic what a permanent-magnet direct current motor is? Then ask them what stator conducting bars are to a rotor causing a rotating magnetic field to generate electromagnetic forces which induce currents in the rotor bars? Also, ask them what an internal permanent magnet synchronous reluctance motor is. Chances are they are going to give you a long blank stare because you just spoke an unfamiliar engineering language to them. ICE auto manufacturers will need to bring in electrical engineers who will need to run parallel to traditional mechanical engineers, or all their current mechanical engineers will need to be also trained as electrical engineers. They will still need to maintain all of their mechanic engineers to service the outgoing ICE vehicles. This applies not only internally in their manufacturing process but also at thousands of franchise dealerships globally.

4) Tesla has something no other Company has at scale – Software engineers, lots of software engineers. Software engineering is deeply embedded into Tesla's S3XY Solar line up extending from their vehicles GUI,

to vehicle battery optimization, to their Solar platforms. These are more key elements in understanding why Tesla has no competition. The future talent and staffing logistics related to EV Mechanic, Software and Solar engineers cannot be understated. Pursuing Trinity will come at a very high cost. It will exact a high financial toll and create many conflicts between new young engineers and legacy talent. This is an absolute massive personnel undertaking having to run two parallel business models integrating differentiating EV, ICE, Software, and Solar engineering expertise's.

5) Tesla also has something no other Company has – the manufacturing, robotics, and engineering Gigafactory platforms to scale its S3XY Solar products. Tesla has robotics where the machines build the machines. The manufacturing genius of its custom Giga Presses (giant custom alloy injected casting machines) cannot be understated. These significantly reduce shipping, manufacturing costs, parts, and assembly times. Tesla leads in all areas. Look for all others to follow and try to emulate. All ICE auto manufacturers will need to run parallel manufacturing floors (One for EVs and one for ICE) or revamp their current ICE platforms to gain efficiencies to drive mass scale EV production. They will need to revamp their entire manufacturing processing and assembly lines to catch the manufacturing speeds and efficiencies of Tesla.

Tesla's ecosystem suite set's it far apart and above from everyone. It's a business model that completely validates

Elon's vision for accelerating the world forward towards Trinity and energy independence. Consumers will adopt Energy Trinity being Triple Hedged Badasses. This global event will be defined as going ludicrous accelerating the world towards energy independence. Strategically, all auto manufacturers will have no other alternatives but to follow in Tesla's footsteps by adopting Trinity. This global event will be defined as plaid accelerating the world forward towards Energy Trinity and energy independence. As Tesla continues to accelerate further away, look for all other auto players to follow them and attempt to mirror the same playbook.

I expect to see significant financial pressure on legacy ICE auto manufacturers due to the aforementioned. Additionally, none of these companies' stocks qualify for Investment Management portfolio ownership under a pure ESG consideration approach. Therefore, any institutional or consumer ownership will be negligible from a growth perspective. I could go on about some even more significant challenges the current EV players and the legacy ICE industry are missing, but I won't. I won't bore you with the many other macro and microscopic issues they face that will give rise to larger issues in the future. Plus, there is no fun in that. Problems and competition spurs innovation and critical thinking. I view strategies, countermeasures, cost savings, opportunities, profitability, and risk differently over a longer time horizon through a different lens. It's also why the entire automotive world doesn't yet realize that they will be directly responsible for plaid accelerating the world towards energy independence sooner than they think.

I believe some automotive and energy experts have claimed that Electric Vehicles will rule the road by 2030. I wish it were true, but simple math points out the error in that claim. Almost 1.5 billion ICE vehicles globally need to be scrapped, recycled, melted down, and re-purposed. If the 2030 statement had merit, it would mean more than 750 million ICE vehicles would have to be dumped and replaced with new Electric Vehicles. Tesla can only produce so many EVs. For the year ended 2021, Tesla produced and sold just under 1,000,000 Electric Vehicles globally. What about adding all the other forthcoming Electric Vehicles by other auto manufacturers? Collectively, they would have to produce 108,000,000+ Electric Vehicles starting in 2023 and repeat this for the next 7 years. Unfortunately, it isn't going to happen.

What about just the U.S.? The U.S. continent is geographically ideally suited for consumers to capitalize on Trinity. The world's automotive players can't deliver at this time, nor will they be able to this decade. There are over 275,000,000 registered ICE automobiles and trucks in the U.S. It's estimated that during 2021, Tesla produced and sold around 350,000 S3XY vehicles in just the U.S. alone (Tesla only provides global sales figures).

Global consumer demand for Tesla's S3XY Solar energy ecosystem and future products will far outstrip production and supply capabilities. I have some recommendations for Tesla on how 2022-2023's production allotment should be handled for fulfillment. Why is this needed? What if Tesla gets 20,000,000 S3XY Solar Energy Trinity orders through next year resulting from this book and the Multiplier model? Unfortunately, Tesla will most likely only have the capacity to produce 1,400,000 vehicles globally for 2022 and maybe 2,000,000 for 2023. Tesla has a future target of building 20,000,000 by 2030. As a result, there will be a massive demand versus production scaling issue that needs to be addressed. Tesla must establish a final order determination for fulfilling 2022 and 2023 deliveries and solar installations. After this, pricing will be on a preliminary basis only and the final cost will be trued up (finalized) at time of delivery. Why preliminary? There will be a multitude of supply chain, inflation and production output delivery dynamics that will be inherently unknown. All these elements weigh in on ensuring Tesla has flexibility in keeping its operating margins in check. Tesla's product demand will explode exponentially. This is just a recommendation on the order of fulfillments:

1) Tesla, SpaceX, The Boring Company, Neuralink, and OpenAI employees. All employees of Tesla and their affiliated companies should have 1st order rights of fulfillment. It is without question that Tesla and their affiliate employees are its most valuable asset. Without them, their product brands would not exist. Order rights would be determined by length of employment service and a limit of 2 Tesla's per employee.

2) Individual Tesla Shareholders should have 2nd order rights of fulfillment. Ownership should afford privileges and shareholders who understood and saw

Elon's vision should be acknowledged for their capital investment efforts. Order rights would be determined by the total number of shares held and the date of ownership. This information would be validated in person by showing their online brokerage positions to a Tesla representative at time of delivery.

3) Current Tesla Owners should have 3rd order rights of fulfillment. Sure, many would view this as unacceptable. They already have a Tesla, or Tesla's so why should they be entitled to more? It's simple. Without their support and confidence in transitioning away from ICE vehicles as early adopters, Tesla would not have the sales, revenues, and related cash flow to be where it is today. Ownership should also convey privileges. Current owners should be rewarded with order rights based on the initial vehicle sales date. For example, those who bought Tesla's in 2012 and are still current Tesla owners have priority over those bought in 2018.

4) Though I don't expect there to be any left over from 2022 – 2023's production, it would be fair game for all other consumers once all the aforementioned fulfillments are handled however, with an added measure. What added measure? Global demand for Tesla's products will far outstrip its manufacturing capability. Growing at exponential rates is extremely capital intensive. In order to enhance its cash flow position and offset the cash deployed related to infrastructure buildouts of its Gigafactory's, Tesla

should begin taking varying levels of advance deposits to determine order of delivery priority. The current $250 deposit just won't cut it. Customers should always be welcomed and allowed to pay in full and upfront for any Electric Vehicle or Solar product in exchange for delivery priority. In other words, consumers that pay in full and upfront for Electric Vehicles and or Solar will have delivery priority over someone who places say a $20,000 deposit. Those that place a $20,000 deposit will have priority over those that place a $10,000 deposit. This might seem totally unfair as it conveys that those that are in a financial position to do so will always have priority over those that do not. This is no different than what has always been witnessed in the housing markets when buyers who have the cash to pay in full typically secure the purchase transaction over those that do not. It is nothing more than exponential demand versus low supply economics. I believe a payment in full strategy will be tolerated. It will be necessary for Tesla to grow exponentially. This will dovetail into Elon's Master Plan: Trinity (Who knows? it could be called Part 3 but Trinity sounds so much better). Such a strategy could turn a negative cash conversion cycle to a positive and exponentially growing cash conversion cycle. Remember demand will be such that disclosures are needed that the final purchase price will be trued up at the time of delivery due to persistent strains on raw materials, manufacturing speed, supply chain logistics, and inflation. Truing up the final sales transaction price will allow for better management of operating margins. I won't comment further on many other ideas I have on Master Plan: Trinity. I do not want to steal Elon's thunder.

Now, don't you wish you were a Tesla employee, a long-term Tesla Shareholder, or were a current Tesla owner? Yes, it's totally fair that Tesla employees, shareholders, and current Tesla owners are afforded rights and privileges for placing their capital and confidence in Tesla's products and services.

Now let's look as some of the EV competition.

21

A LOOK AT SOME CURRENT ELECTRIC VEHICLES

"If you're entering anything where there's an existing marketplace, against large, entrenched competitors, then your product or service needs to be much better than theirs."
– Elon Musk

The competition should never forget they are and will be following Tesla's playbook. Needless to say, Elon and Tesla stand far and above the competition based on their mission that started almost two decades ago. It's what separates Elon and Tesla from all the competition: Leading the way forward to save Earth and the future of Humanity through sustainable energy and global consumer energy independence through their S3XY Solar vertically integrated energy ecosystem.

These noble efforts by Elon and Tesla along with their state-of-the-art product offerings are why they have so many ardent fans and supporters across the world. Elon's Secret Master Plans and Tesla's S3XY Solar Energy ecosystem led me to an epiphany in formulating everything discussed and presented in this book. Without them, I wouldn't have put forth the effort to bring The Energy Matrix, Trinity, being Triple Hedged, and the Multiplier to light nor dig deeper and deeper to discover additional X-Factors.

It will most certainly be challenging times in the automotive industry these next few decades. Nonetheless, they should all take comfort knowing they will be directly participating in completing Elon's vision in accelerating the world towards sustainability through energy independence. This will ensure a healthier Earth for generations to come.

Here are a couple of observations on the current competition:

Audi e-tron – Moving almost 3 tons of mass for a top speed of 124mph on a 95 kwh battery pack results in a not-so-efficient energy delivery system. With an EPA range of only 222 miles owners will be dismayed at how frequent they need to recharge and wonder why did they just drop $70,000. How about the $45,000 Q4 e-tron? A top speed of 99mph with a 7.9 second 0-60 time may make an owner feel like the vehicle is always running low on a state of charge.

Chevrolet Bolt EV – GM recalled all 141,000 Bolt EVs in August 2021 due to known fire outbreaks. On top of recalling the vehicles and replacing the battery modules, GM pushed software updates limiting the battery's state of charge so as not to go over 90% or go below 70 miles. Additionally, it asked owners not to charge overnight or park their cars in garages while charging for risk of fire. The recall is expected to take months as it receives new batteries from

LG. Will these Bolt owners ever buy another GM product? I encourage you to just ask one. There were only 358 Chevy Bolts sold during the entire first quarter of 2022.

Ford F-150 Lightning – Ford stuck with the design queues of the legacy ICE F-150 both on the exterior and interior. Smart play. American's love their Ford F-150's as evidenced in their sales figures. Love the ginormous frunk and number of power outlets.

Ford Mach-E – Unbridled? Step into the Mach-E and tell your passengers they are about to experience unbridled mode? That's a head-scratcher. Why not leverage off the name? Sounds way cooler to tell passengers they are about to experience "Mach Speed!" It's an easy software update. Hopefully, Ford will also add perimeter cameras and fix the center screen. It needs to be horizontal and not vertical (consumers who already bought Mach-E's will come to find out why later).

GMC Hummer EV – At a starting price tag of $110,295 and a 212-kWh battery pack it further demonstrates that the GM legacy lunacy lives on. It has zero clue on affordability let alone profitability in building a vehicle designed for .000000000001% of the population. Sorry engineers and manufacturing teams, I know you had your marching orders. Enough said.

Hyundai Kona Electric – 0-60 in 7.9 seconds is slow motion in the EV world. Combine that with an engineering design flaw with the charging placed in the front bumper.

According to a 2019 crash statistics report from the National Highway Traffic Safety Administration, 54.2% of passenger vehicle crashes by initial point of contact was damage to the front end[7]. Ask any former Nissan Leaf owner why they wouldn't recommend putting the charge point in the front bumper of the car.

Hyundai IONIC 5 – No sizeable frunk and sub trunk. Just one engineering flaw: Steering paddle-controlled regenerative braking. This is another head-scratcher as constant and continuous added input is needed to use/control/change the brake regeneration.

Jaguar I-PACE – Crowned as the first "Tesla killer" Electric Vehicle. In 2019, the I-PACE won World Car of the Year, World Car Design of the Year and World Green Car. These awards were by a panel of 86 motoring journalists from 24 countries and was won just weeks after winning the European Car of the Year title. Surely, such awards by so many accredited motoring journalists and countries would elevate the I-PACE's status as the most desirable electric car in the world? Unfortunately, the following global sales numbers for the I-PACE suggest otherwise:

- YE 2019: 17,355
- YE 2020: 16,457
- YE 2021: 9,970
- 1Q2022: 2,014

The Tesla Model Y, which launched in 2020, sold over

---

[7] https://crashstats.nhtsa.dot.gov/Api/Public/ViewPublication/813141 p.98

500,000 globally in just two years:

- YE 2020: 84,160
- YE 2021: 420,618

Sadly, the sales data reflects what consumers genuinely want despite the number of awards and accolades the I-PACE has won. As such, I do not believe that the I-PACE is destined to be a "Tesla killer" anytime soon. Jaguar needs to ask consumers why they chose a Tesla over an I-PACE. Once they understand all the issues related to their I-PACE, they will have a starting point on how to start turning things around.

Kia EV6 – Same engineering flaw as the Hyundai IONIC 5: Steering paddle controlled regenerative braking and no sizeable frunk and sub trunk.

Kia Nero – Sadly, double engineering flaws – Front bumper charging and steering paddle controlled regenerative braking.

Lucid Air – A Saudi Arabia owned and controlled Electric Vehicle company fit for Royalty. Posh, excellent performance, and engineering specs along with being touted by Automotive magazines and YouTube Influencers as a "Tesla killer." Sadly, Lucid's own self-business dealings may end up being a "Lucid killer" to potential customers from an environmental risk approach. Lucid, based in California, is 62.7% controlled by Saudi Arabia, an Oil Cartel member. California is a state known for progressive environmental policies. Wrap your head around that. According to Lucid's August 20, 2021, amended SEC Form S-1 Page 126, Ayar Third Investment Company, a wholly owned subsidiary of The Public Investment Fund, owns over 1 billion shares of LUCID, equal to 62.7% ownership.[8] Ayar is a wholly-owned subsidiary of the Public Investment Fund, which is the sovereign wealth fund of the Kingdom of Saudi Arabia. The Board of Directors Chairman of the Public Investment Fund consists of His Royal Highness Prince Mohammad bin Salman Al-Saud. How much control does Saudi Arabia exert over Lucid? On February 28, 2022, Lucid announced in a Press Release[9] that its first international plant will be in Saudi Arabia, hence an EV fit for Royalty. It states, "This new international manufacturing plant targets 150,000 vehicles per year at the King Abdullah Economic City and will leverage U.S. engineering, R&D, and manufacturing expertise in assembling new models for global markets." This strategy seems odd from a global markets sell approach. If one looks at where Saudi Arabia is geographically in relation to where global market demand might be, say Europe and Asia. This strategy does not make business or environmental sense. Just look at a map and see the shipping transportation logistics, added shipping costs and the resulting pollution created to ship these 150,000 vehicles to Europe and Asia. Recall that Saudi Arabia owns 94% of Amarco. Amarco remains the single greatest contributor to global carbon

[8]https://www.sec.gov/ix?doc=/Archives/edgar/data/0001811210/0001104659 21107865/lcid-20210819xs1a.htm
[9] https://www.lucidmotors.com/media-room/lucid-group-gearing-up-first-international-plant-saudi-arabia

emissions of any company in the world since 1965. How will automotive and news journalists report on Lucid being owned and controlled by Saudi Arabia and reconcile its ownership of Amarco being directly associated with the single greatest contributor to global carbon emissions? How will the Oil Cartel stigma and massive global emissions factor weigh on the minds of Earth eco-conscience Lucid owners? Additionally, look at the media's recent criticism of Phil Mickelson and others for joining the Saudi-backed LIV Golf Series, Saudi risk dynamics tied to the terrorist attacks of September 11, 2001, and assassination of Jamal Khashoggi. Unfortunately, the aforementioned makes Lucid's stock the least likely of any auto manufacturer to be held by Environmental, Social, and Governance conscience Investment Management Companies and individuals.

Mini Coupe Electric – With a max range of 114 miles per full charge, it's not an option for those looking to travel across the U.S. It's more ideally suited for consumers in dense urban cities who only have very short commutes.

Nissan Leaf – Sales were never great on the Leaf. Sadly, it continues on that trend after over a decade in production. Nissan should deeply explore why this product missed the mark and continues to do so, as evidenced by the long dark shadows cast by all the Tesla's zooming right by them.

Polestar 2 – A Chinese-built electric vehicle disguised with a Swedish Volvo affiliation. The Polestar 2 is built in Chengdu, China. Volvo partnered with Geely, a Chinese-based auto manufacturer, to create Polestar. I believe most U.S. consumers would be apprehensive against buying and driving a Chinese-built vehicle from a safety, reliability, manufacturing, and technology perspective. It's untested and unproven in the U.S. market. They face the same challenges that Honda and Toyota did entering the U.S. market in the 1970s and Hyundai did in the 1990s.

On April 4, 2022, Hertz inked a deal with Polestar in the U.S. It's a questionable strategic decision for the U.S. market coming off Covid-19 combined with even more risk exposure dynamics related to China. Polestar would fare better in China and other countries.

Porsche Taycan – An EV debut that was on-point with the looks and performance to carry on the Porsche design and racing heritage. I don't think they are building to scale for vertical growth. There are many pricey customizable options to choose from that add up quickly. Who knows? Maybe they want their brand to remain a small-scale, exclusive boutique EV that remains just out of reach from a majority of consumers?

Rivian – Being first is not always a good thing. Also, being bigger is not always better either. Case in point is Rivian's new R1T. It weighs a staggering 7,148 lbs. and is classified by the EPA as a Class 2b heavy-duty truck. This translates into an EPA of 48kwh/100 miles and an MPGe of 74/66MPGe. An R1T with a large 128.9 kWh battery pack with an EPA range of 314 miles was put through a 75mph real-world driving test. It's touted 314 miles of range only delivered 220 miles, translating into a real-world efficiency of only 35 MPGe.   For comparison purposes, a Tesla Performance Model Y yields about the same 300 mileage range using only a 75-kWh battery pack. Yes, they each serve a different utility purpose however, Rivian's needs 67% more in batteries/energy to travel the same distance. This means longer and more frequent recharging coupled with higher recharging costs.

I also believe Rivian's engineers made a design flaw by placing the charger in the front driver's side front bumper. According to a 2019 crash statistics report from the National Highway Traffic Safety Administration, 40.3% of large truck vehicle crashes by initial point of contact was damage to the front end[10]. Hopefully, it's an easy engineering fix to relocate to a safer location.

Lastly, in July 2020, Tesla filed a lawsuit against Rivian alleging that Rivian stole trade secrets from them by actively recruiting former Tesla employees and instructing them to provide specific proprietary confidential and sensitive information. If it turns out to be true, it will be a public relations and shareholder nightmare. It could be financially devastating to Rivian for theft of intellectual property.

Volkswagen iD3 and iD4 – Sadly, lacking power and no frunk. Musk's muted reaction when driving it said it all when Herbert Deiss, now former CEO of VW, eagerly asked him to test drive the iD3. Any VW owner, along with Tesla owners

---

[10] https://crashstats.nhtsa.dot.gov/Api/Public/ViewPublication/813141: p. 102

would also most likely agree - Nein farfegnugen! Sales are good for now, however, how long will consumers be satisfied with nein farfegnugen?

Volvo C40 and XC40 Recharge – Some are sourced in Ghent, Belgium, others from China. As noted with Polestar, I believe Chinese manufactured vehicles for sale in the U.S. is a very risky play.

Do your own research, schedule test drives, and talk to as many Electric Vehicle owners as possible when you go Electric. Apart from Tesla, many EVs are new and have yet to be tested over the long term. This poses reliability risk to new Electric Vehicles. Every driver has a different need for their driving utility requirements. An Electric Vehicle is a large financial investment. You want to make the right financial decision on your first Electric Vehicle and without buyer's remorse (case in point, Chevrolet Bolt EV buyers).

ICE and other Electric Vehicle owners will eventually come to understand why Tesla's S3XY Solar ecosystem stands alone. This is evident in their global booming sales figures and demand across all product lines. I dare say, Tesla is currently the most iconic, most talked about, and most exciting brand in the world. There is a genuine passion and excitement for their products collectively demonstrating why the next generation of driver's want Tesla's. How much passion and excitement? To date, Tesla has generated $200 billion in global revenue since its formation in 2006 yet, has not spent a single dollar on advertising. This is unprecedented with any global brand. 16 years of zero advertising! No Superbowl® or television commercials. No paid Celebrity or YouTube endorsements.

No global company, not Apple, not Amazon, not Google, not Microsoft, not Coca Cola, not Disney, not Nike, not Walmart, and not even Louis Vuitton can lay claim to this. These global brands spend billions every year to keep their products in front of global consumers. Such unprecedented global brand strength demonstrates that Tesla's suite of products are consumer driven, ultra-high end, innovative, and exciting products that sell themselves.

The safety, engineering, technology, performance, styling, reliability, sales delivery experience, mobile on-site service, global network of growing Superchargers, and perpetual drive to challenge, optimize, and refine every aspect of transportation and energy dynamics clearly shows the strength of their product line up. Most consumers will come to find if they talk to any Tesla owner, they will most likely talk your ear off. If you are in the market for an Electric Vehicle, I would encourage you to pull into any Tesla Supercharging station for an opportunity to talk to multiple owners regarding their driving and ownership experience. You will come to understand the level of passion and excitement Tesla owners have for their products and experiences.

I will share with you our family experiences on why our Tesla's elicit joy while hopefully, not talking your ear off.

22

JOY MULTIPLIER™ +
FREE ENERGY MULTIPLIER™

"Great companies are built on great products." – Elon Musk

I wrote this to a family member back in mid-2019 who asked how we liked our Tesla's. Some things have changed over the last three years however, much remains on-point:

Hi XXXXX,

Seeing it, touching it, interacting with it, driving it, accelerating in it, braking with it, washing it, cleaning it, traveling in it, playing with it, and charging it – Our family's three Tesla Model 3s all elicit joy. Let me clarify on eliciting joy. Well, you've heard of Marie Kondo's phrase, Spark Joy®? This occurs when you touch an object and it lifts your body spirits with a wonderful goosebump sensation and gives you the warm and fuzzies. Well, Elon Musk and Tesla went far and above sparking joy for our family. How much joy? Our family's Tesla's are a **Joy Multiplier™**, to be specific. A Joy Multiplier is being surrounded and immersed by many joyous things. Yes, that much joy!

We can all tell you without a doubt that our Tesla Model 3s are the finest vehicles we have ever owned and driven! We've owned Acura's, Lexus', Honda's, and Toyota's. Everything about the vehicle is revolutionary and is a disrupter to the automotive world. I would say the experience parallels similarities to when people of a hundred years ago went from the horse and carriage to the automobile. They didn't look back. Back then, people were so used to their old faithful horse Becky and their carriage that they couldn't possibly fathom a motorized contraption replacing their reliable horse. Some may say Tesla is nothing more than an expensive, souped-up golf cart with a tablet for instrumentation. I see it as the safest, most environmentally efficient means of transportation infused with state-of-the-art technology and engineering encased in a masterfully executed exterior design imbued with a Zen interior. Our Tesla's hit all the marks for us. Think of Tesla's as the latest iPhone or Samsung Galaxy – they can do so much with so little.

The Model 3s ride experience is like that of a Porsche Carrera or BMW M3. Tight and taught. The steering is spot on and the low center of gravity floor placed battery pack has the car planted to the ground. Couple that with instantaneous, linear torque throughout the acceleration band and you have an E-ticket ride at Disneyland - the Matterhorn and Space Mountain! Do you get the picture? Smiles and joy abound. It's an E-ticket ride every time we drive it. These are just some of the joy multipliers. There are plenty more.

The center-mounted screen was at first knocked by the automotive world, but it too is a well-executed engineering masterpiece. The thought from the engineering to form and function is absolutely brilliant and way forward-thinking. Yes, it took a day to get use to glance off to the side but just like when smartphones first came out, consumers were quick to adapt due to the interface being straightforward and intuitive. Because the large center screen pad is software driven, it can be continuously enhanced over the car's lifetime. When you buy a traditional car, nothing changes as the instrumentation stays the same forever.

Our Tesla's can be continuously updated and optimized with software over-the-air updates as we have experienced with ownership. Want an extra 5% power for quicker acceleration? Done – with a software update. Want to watch Netflix or Hulu on the 15-inch center mounted screen? Done – with a software update. Unlock battery reserve to increase driving range? Done – with a software update. Want the Tesla to make funny fart noises? Done – with a software update. Want to keep the air conditioning on for your dog and display a message on the screen reading the air condition is on at a cool 70 and my owner will be back shortly with an animated dog wagging its tail on the screen? Done – with a software update. Want to play chess or play video games from the screen? Done – with a software update. Want to get frisky in the car creating the next generation of humanity? All you need to do is engage Romance mode. Done – with a software update. The center-mounted software driven screen by Tesla is pure genius as it allows for the Tesla to be continuously refreshed and be brought up to date without the need of going into a service center with all software updates performed over the air.

Tesla is continuously pushing down software updates. To quote an owner who was interviewed about Tesla regarding their center-mounted screen and interface, he said, "It's like Christmas all of the time with a Tesla. There are always enhancements and updates being rolled out. It's exciting getting into your car and seeing what new presents are awaiting you." We can absolutely attest to this. No other car has Dog Mode to ensure a cool temperature for our pooches. No other car has Romance Mode... ;-) ABSOLUTELY LOVE IT! Sentry mode protects the vehicle and records footage from its eight exterior cameras providing 360-degree viewing. Our Tesla's horizontal mounted screen is a full-on entertainment system from movies, to music, to arcade games, to the internet. These are fantastic must have features to have while supercharging. We just received updates this week for Caraoke to sing Karaoke in the car! We can't wait to see what future ideas and enhancements will be forthcoming!

One of the main reasons why we got our Tesla's is safety. Our lives are absolutely worth it. The Model 3s debut scored the highest ever safety ratings from both the National Highway Traffic Safety Administration in America and the Euro New Car Assessment Program. Tesla's revolutionary design and safety engineering disrupted hundreds of millions spent by ICE legacy auto companies on computer-animated design engineering for their legacy platforms. Tesla's vehicle platform engineering represents a giant leap in automotive safety over traditional ICE vehicles.

We have had no problems or concerns with our Tesla's whatsoever. We have not had to take them in for any type of maintenance. The only thing we have had to do is add air in the tires, add windshield wiper fluid, and change out the internal air cabin filters.

On charging at home - we installed two, 14-50 NEMA 240-volt outlets in our garage. This is the same type of electrical outlet for electric dryers or RV camper hookups. It allows us to conveniently charge our Tesla's at a rate of 30 miles per hour. It is possible to charge from a regular household 110-volt outlet. However, it will only charge at a rate of 4-5 miles an hour. We absolutely love the convenience of charging at home! We love never having to visit a gas station again waiting in the long Costco gas lines! We can have a full driving range of 300 miles every day if we want to with the convenience of charging at home.

What are some of the other things that are joy multipliers? A lot more! No oil changes forever! Unlike gasoline cars, Tesla cars require no traditional oil changes, fuel filters, spark plug replacements or the risk possibility of stolen catalytic converters. No more smog checks forever! We may never need to install new brake pads because regenerative braking returns what would normally be wasted energy in an ICE car, back to recharge the battery. This significantly reduces or eliminates brake wear when done right. It brings the car to a complete stop and learning to judge the distances by modulating your foot off the accelerator is easy.

The battery range of 325 miles is the best out there and it gets us through without having any sort of range anxiety, most often expressed by gas drivers. We charge at the convenience at home in our garage, which is 99% of the time. We love the growing network of Superchargers and the speed at which they charge (now up to 1,000 miles per hour - so a full charge in about 15-20 minutes). All other electric vehicles are geographically bound, meaning if you buy anything other than a Tesla, you cannot really go on a long-distance commute due to the length of time it will take at other EV charger stations which are minimal. Want to travel across the country in a BMW i3, Nissan Leaf, Chevy Bolt, or even with the new Jaguar I-PACE? Good luck at finding a widely available charger. Tesla's supercharger network is proprietary and only for Tesla's in the U.S. No other EV's can charge at Tesla's Superchargers. Tesla's Supercharger network is brilliant and will get you to and from your destination anywhere across the U.S. We always felt guilty driving around with a CO_2 ICE emitting vehicle and now our conscience is fully at ease with our Tesla's since they are primarily recharged by our homes Solar Energy System.

I would say Tesla is pretty much where Apple was when Apple introduced the iPhone. Sure, there were other phones made by experienced phone companies (BlackBerry, Nokia, Ericsson). Then Apple came along with the iPhone and it totally transformed the user experience. Fast forward over a decade later. Now, it's only Apple and Samsung that dominate smartphones. I would draw the same kind of analogy with Tesla to the rest of the automotive world. Tesla has re-invented the automobile. Just ask the next generation of any young driver. A majority of them all want Tesla's.

A hundred years ago, people were so used to the horse and carriage as a means of transportation. They scoffed and laughed at a mechanical horse (now the current automobile). We are witnessing the next evolution stage of the automobile. Gasoline engines are very inefficient when compared to a battery-powered magnetic induction drivetrain engineered and mastered by Tesla.

We had Solar installed, so it's the equivalent of having our own gas station and our own electric power station. It recharges our vehicles and offsets all our household electricity expense.

15 years ago, people laughed at Amazon for online shopping just like they scoffed at Apple for coming out with their iPhone and a touch screen keyboard. Today, they have captured and won over the consumer's hearts. I believe Tesla is following in their footsteps in capturing and winning over global consumers. They have certainly won us over with all the many joy multipliers!

So far, Tesla has been delivering exceptional customer service. By this, I mean being anticipatory of a client's needs - giving them stellar service before they even request it, listening and having the answers to their questions, going above and beyond their expectations, being solutions-oriented, calling them by name and thanking them for their business. It seems straightforward, but so many companies fail to get it right.

Tesla has re-imagined the customer experience from buying the car online to servicing the vehicle through mobile on-site servicing. What you see is what you pay. There is no salesperson or sales manager to haggle with.

Tesla is delivering on giving the customer what they don't even know they want and once they have a Tesla, they will most likely never give it up. These are just some of the many reasons we are sold on Tesla and how their exceptional customer service is hitting all the marks on the engagement end.

We are very, very, very happy with our Tesla's. The design, aesthetics, technology, maintenance, safety, handling, performance, and environmental benefits are all truly joy multipliers!

The aforementioned was my story about how well we liked our Tesla's back in mid-2019. Post thoughts since 2019? Our Tesla's continue to impress and deliver on the excitement and the joy multipliers. Still zero maintenance, constant over the air updates with new games, features, and entertainment. Tesla continues to stand alone with its proven design safety, battery and software technologies, AC induction, and permanent-magnet direct current propulsion architectures, handling, and performance, and almost 30 billion miles driven collectively by global Tesla owners complemented with a robust and growing global infrastructure charging network. This past holiday in December 2021, Tesla pushed down the Tesla Light Show to our Model 3s, a testament to the presents that just keep coming. Yes, everything remains spot-on, however, with yet another (you guessed it) new multiplier discovered.

What are we happiest about the most with our Tesla's? Knowing that all three of our Tesla's and our Solar Energy System will be fully cost recovered by eliminating our electricity and gasoline expenses, making them all free. The Multiplier model I built back in late 2018 demonstrated this and projected the financial outcome. To date, we have fully cost recovered one Tesla through the elimination of recurring gasoline and electricity costs. We expect to recover all our remaining costs by the year 2031 by being energy positive, empowered as our own utility.

Our Joy Multipliers have a new best friend. It is said that some of the best things in life are free. By decoding two key fundamentals of Elon's Master Plans and unlocking the "X" Multiplier, our Tesla's, household electricity, and Solar are free compliments of the free energy from the Sun. I collectively refer to the symbiotic relationships as a **Free Energy Multiplier**™ - Free Energy from the Sun, free Solar, free Electricity, free Electric Vehicle's, all free of CO_2 emissions. Elon's Secret Master Plans imbue the beauty of these synergies: Free energy from the Sun to energize Solar into free energy to power consumer's households and businesses and their Electric Vehicles all translating into symbiotic transportation and energy dynamics free of CO_2 emissions. Joy multipliers abound plenty. Now do you understand why I coined the phrases Joy Multiplier + Free Energy Multiplier? I hope my attempts to demonstrate it with my Multiplier model and this book resonate true to that.

If your circumstances model out not being able to take full advantage of the Multiplier by not being fully energy positive, you can still take comfort in knowing you are still realizing cost recovery savings through Solar cost offsetting. This stands far above any ICE strategy in not being able to recover any costs. Some is better than none, but free is always better!

Tesla's ecosystem comprised of its S3XY Solar solutions is the only company vertically positioned to allow consumers to capitalize on free! Remember, always capitalize on free! Never waste free!

23

ELON'S
IMMORTAL MAGNUM OPUS™

"Tesla is to protect life on Earth, SpaceX to extend life beyond."
– Elon Musk

Magnum Opus represents the greatest achievement of an individual. I coined the phrase, **Immortal Magnum Opus™** because like with being Triple Hedged, it's just downright badass. Decoding Elon's Secret Master Plans along with the development of the Electric Vehicle and Solar Savings Multiplier unlocked secrets into how far-reaching Elon's vision, strategies, and accomplishments will forever change the direction of Nature, Humankind, and Earth – An Immortal Magnum Opus.

In short (or in this case, long), we come to unlock the following multidimensional facets by decoding just two key elements of Elon Musk's Secret Master Plans:

- Elon Musk is executing and seeing through on his Immortal Magnum Opus. Trinity (Electric Vehicles, Solar, and Battery Storage) was executed on directly as a result of Elon's epic, forward-thinking vision, strategies, determination, and accomplishments.

- Consumers will awaken within the Energy Matrix realizing the future financial horrors of being subject to the Energy Dependent Monopoly Model. Consumers will understand why the combination of Electric Vehicles and Solar are a winning financial strategy.

- Consumers will come to fully understand that the Sun's energy is 100% free, widely abundant, and greets the world at every sunrise. Consumers won't waste it.

- Consumers who strategically transition to Electric Vehicles paired with Solar to power their transportation and electricity needs eliminate three long-term risks: fossil-fuel dependencies, electricity dependencies, and CO_2 emissions. By doing so, they will be positioned for both long-term financial and environmental success. They will become energy independent triple hedged Energy Trinity bad asses.

- Consumers will come to understand the simplistic beauty that the Multiplier reveals: The more Electric Vehicles one acquires, the more one saves due to the higher cost structure of gasoline relative to the nominal incremental amount of solar energy needed to eliminate up to 40 years of both gasoline and electricity costs.

- Consumers will pursue the Free Energy Multiplier model – Free energy from the Sun, free Solar, free Electricity for their households and businesses, and free Electric Vehicle's.

- Tesla has no competition. It leads in all areas, delivers on the joy multipliers, and continues to accelerate away from the competition. Every auto manufacturer will follow in their footsteps and attempt to mirror their playbook.

- Tesla's vertically integrated S3XY Solar energy ecosystem accelerates the world's transition towards global energy independence. Consumers will become energy positive, empowering themselves as their own utility. This will catalyze a global movement leading to ludicrous acceleration towards energy independence. Global auto manufacturers will copy Elon's playbook and go full-blown Energy Trinity leading to plaid acceleration of the world towards energy independence. Each event will sequentially culminate in the realization of Elon's Secret Master Plans and overarching goal – An entire global movement that accelerates the world forward in sustainable energy and consumer energy independence.

- All of the aforementioned transcends Elon Musk's Immortal Magnum Opus – His overarching vision, strategies, determination, accomplishments, and open patents have and will forever bring about a better and brighter future for Nature, Humanity, and Earth.

Please, do tell everyone.

24

CONCLUSION

"If you want the future to be good, you must make it so. Take action to make it good, and it will be." – Elon Musk

Thank you for allowing me the opportunity to free your mind to the financial horrors and realities of The Dependent Monopoly Model. My goal was to awaken you to a new energy paradigm created by Elon Musk and Tesla, along with just a few of the key benefits their multipronged Energy Trinity strategy and product offerings afford.

Elon put out his Secret Master Plans waiting for someone to come along and decode them. I hope my attempts to decode just two key aspects within his Secret Master plans: 1) Becoming energy positive 2) Empowering oneself as their own utility and unlocking the Multiplier did them justice.

My goal at the very beginning was five-fold. I hope I accomplished this and much more. A sixth one was added as a little bonus from a strategic investment play perspective as follows:

1) You will be thanking Elon Musk and Tesla for executing on his Secret Master Plans.

2) You now understand why Electric Vehicles and Solar are a winning financial strategy.

3) You will come to understand that the more Electric Vehicles one acquires, the more one saves due to the higher cost structure of gasoline relative to the nominal incremental amount of solar energy needed to eliminate both gasoline and electricity costs.

4) You will gain insight into how I unlocked and decoded two key elements within Elon's Secret Master Plans.

5) You will evaluate and implement your own financial strategy by becoming energy positive, empowered as your own utility. In doing so, you may get a free Tesla and Solar Energy System or will at least recover a good portion of your costs.

6) You will benefit both in knowledge and financially from my TESLAMAXSFAST stock picks.

I'm executing and seeing through on my own Magnum Opus: 1) Educating consumers with this book. 2) Educating consumers with my dynamic Multiplier model. 3) A goal of saving consumers trillions in future gasoline and electricity energy costs. 4) As a result of the aforementioned, save Nature and Earth exponentially through the elimination of carbon emissions 5) Educate and enrich consumers with my TESLAMAXSFAST formulated investment strategy.

My post PCBS leadership conflict provided me with new perspective and direction. None of this would be possible without Elon's Secret Master Plans, Tesla employees executing on these plans, Pacific Coast Banking School, and the loving support of my family. Building the Multiplier models, writing and self-publishing this book, formulating all of the frameworks coming out of The Energy Matrix, Trinity, being Triple Hedged, Elon's Immortal Magnum Opus, designing the book cover art, designing, and building the Energy Trinity website, along with formulating my TESLAMAXSFAST strategy for the benefit of all consumers provided me with an opportunity to demonstrate my ongoing leadership and development. Steve Jobs was right. I trusted that my life's dots would connect in my future. I trusted my gut, destiny, life, and karma. These life dots gave me the confidence to follow my heart. I hope everything I developed in this book will both enrich and inspire others. I can say with much satisfaction, this journey and experience has been yet another love and joy multiplier in making my life better and brighter.

It is my hope that the aforementioned will empower you as a leader to guide your families and communities towards both financial and energy independence. Together, we can quickly lead with courage, determination, and knowledge in setting an example for the rest of the world to follow by leaving a positive and impactful legacy. By doing so, we can all reduce our carbon footprint for the benefit of Nature, Humanity, and Earth.

I cordially welcome and invite you to join Elon and Tesla on the very mission they started on almost two decades ago: To save Nature, Earth, and the future of Humanity by transitioning the world to Trinity and energy independence through their S3XY Solar energy product solutions. Elon's overarching mission, Secret Master Plans, and Tesla's vertically integrated energy ecosystem is a combination to which no other auto manufacturer or energy company can lay claim to. Your patronage of their products and services will ensure both Tesla's and Earth's long-term success and prosperity.

We are at a critical time and inflection point wherein consumers can now change the course of history due to a new energy paradigm created by Elon Musk and Tesla. I encourage you to cement your legacy in being a part of Tesla's global transportation and energy revolution. Tell future generations when you awakened within The Energy Matrix and that you took part in Tesla's S3XY Solar Energy revolution for the greater good of Nature, Humanity, and Earth to ensure a better and brighter future for many generations to come.

I will conclude this book the way I started it with parallels between the realities of The Energy Matrix in relation to the science fiction thriller, *The Matrix*. Agent Smith witnessed human interactions and development through various observations immersed in multiple simulations within the Matrix. He concluded that Humans aren't actually Mammals. They are in fact, a disease, a virus, a cancer, and a plague upon Earth. Smith observed and clearly delineates that every Mammal on Earth instinctively develops a natural equilibrium with their surrounding environment, whereas Humans do not. Humans he noted, mimic a virus, living freely off its Earth host, consuming every natural resource, and multiply until those resources are exhausted before spreading to another area and repeating. The Energy Matrix, Trinity, and being Triple Hedged shatters this notion as a result of this new energy paradigm created by Elon Musk and Tesla. By decoding Elon Musk's Secret Master Plans, Humankind is now better positioned to act in harmony restoring a natural equilibrium with Nature and Earth.

Do all my multidimensional dynamics in formulating Elon's Immortal Magnum Opus and the aforementioned give rise to multiple financial and environmental Black Swans, both favorable and unfavorable? That is for Humankind to decide and act upon. The industrial revolution periods were followed by the current technological revolution. Let it be known that Tesla started the Trinity revolution. A new energy paradigm that leverages energy symbiotics and energy independence that encompasses both global transportation and consumer electricity energy dynamics. Humankind needs to find the strength, leadership, and courage to develop a natural equilibrium with their surrounding environments. Choose solutions that lead to Nature's, Earth's, and Humankind's prosperity and evolution? Or remain on its current path that leads to its demise and extinction? The future of Nature, Earth, and Humankind is upon all of us right here, right now. Everything collectively designed in this book was provided to be fact, solutions, and inspirations based to encourage and empower you to free your mind. You have been awakened within The Energy Matrix and are no longer subject to being dependent on the Monopolies.

As of the date of this book, September 2022, Tesla is

offering consumers the ability to earn a $300 rebate award for either a Solar Roof or for Solar Panels (upon installation and permission to operate). If you are interested in Tesla's Solar Energy solutions and you have a family member or friend with a Tesla, please use their Tesla referral link so they may get credit for it. However, if you do not have a family member or friend who owns a Tesla and you found this book and the Multiplier beneficial, I humbly ask that you please use my referral code:

https://www.tesla.com/referral/frank86709

I humbly and graciously thank you for all your time and consideration. I wish you all the best in your life's pursuit of a life of love without limits and a life lived without regrets, many Love and Joy Multipliers, Energy Trinity, Free Energy and Environmental Multipliers, and a life of natural equilibrium with Nature and Earth by being a Triple Hedged Badass!

Please do me a huge favor and leave your feedback regarding this book. It gives me great motivation and direction when I hear positive feedback. It may not address all your questions but rest assured, I am always solutions and ideas oriented. I will use your comments objectively to help me grow and develop in areas that I need to continue to improve upon in formulating and delivering solutions. Also, if you found this book beneficial, please help me. Tell family, friends, and colleagues about purchasing their own book copy (preferably an ebook copy — it's eco-friendly and Mother Earth approved). The ebook copy is available on Barnes and Noble, Apple Books and Google Play. I need your assistance and help in getting the word out in sharing this book through your leadership, educational, media, and social media channels.

Mother Earth is unyielding in the life, love, and many wonders she gives day after day, month after month, year after year to her 8 billion children. Her 8 billion children are straining and stressing her through global warming caused by high levels of CO_2 emissions. Let your voices and actions be known that the quicker consumers stop ingesting black pills and escape the Dependent Monopoly Model, the quicker we can restore a natural equilibrium with our surrounding environments. Let your voices and actions demonstrate that Humankind can love Mother Earth as much as Mother Earth so graciously produces and provides for the benefit of Humankind.

Let us not be mired in the problems that lie before us but be strategic, innovative, and resolute in formulating solutions that lead to Humankind's, Nature's, and Mother Earth's prosperity in pursuing and accelerating Trinity. Elon's overarching Secret Master Plans, vision, strategies, determination, accomplishments, and open patents have given the world Trinity – An Immortal Magnum Opus that transcends space and time.

This is Humankinds entrance into a new Golden Age that leverages technology and energy symbiotics of both consumer transportation and energy dynamics. I call it the Golden Age of Trinity. This is our time. Un Giorno Per Noi. (This is your reminder to now look up the Italian to English lyrics translation). A time for us. A time to act now for the future benefit of Nature and Humanity. A time for Trinity. A time for Mother Earth. This is our time to show Mother Earth a life of love without limits and a life lived without regrets.

Start ordering your Tesla S3XY Solar solutions. I have a little hunch future demand will be just a whee more elevated. ;-)

Remember, energy power to the people, for the people, and by the people as only Elon, Tesla, and his Secret Master Plans can!

Sol ombibus lucet.

WEBSITE

I built and designed the ENERGY TRINITY™ website, logo, other trademarked logos, phrases, and incorporated design elements in relation to Tesla, the Model 3, and Elon's Immortal Magnum Opus. The Model 3 is a key defining invention and engineering masterpiece that marks a precipice moment and turning point in the future of Humankind – the first mass produced consumer Electric Vehicle that disrupted multiple industries and ideologies around the world.

You will see parallel correlations in the design queues in the letter "E's" from this book in the word ENERGY. The first "E" is vibrant green and represents the green jelly bean, Tesla's green energy Model 3 and other energy products, Earth's vibrant green tundra, and money savings. The second "E" reflects Elon's vision for a crystal blue Earth and atmosphere. The word TRINITY captures these symbolisms with the base of the word reflecting Earth's vibrant green tundra transitioning and leading up to a crystal blue Earth and atmosphere – the overarching goals of Elon Musk. His Secret Master Plans give rise to an Immortal Magnum Opus that transcends space and time that will forever change the course of history for a better and brighter future for Humankind, Nature, and Earth.

Collectively, Elon Musk and his Secret Master Plans gave Humankind new technology, energy, and transportation alternatives that work together symbiotically and in harmony to lead us into the Golden Age of Trinity.

www.energytrinity.com

www.energytrinity.com

Made in the USA
Monee, IL
26 November 2024

71247511R20164